Pro Tools LE and M-Power

Pro Tools LE and M-Powered
The complete guide

Mike Collins

ELSEVIER

AMSTERDAM • BOSTON • HEIDELBERG • LONDON • NEW YORK • OXFORD
PARIS • SAN DIEGO • SAN FRANCISCO • SINGAPORE • SYDNEY • TOKYO
Focal Press is an imprint of Elsevier

Focal
Press

Focal Press is an imprint of Elsevier
Linacre House, Jordan Hill, Oxford OX2 8DP
30 Corporate Drive, Suite 400, Burlington, MA 01803

First published 2006

Notice

No responsibility is assumed by the publisher for any injury and/or damage to persons or property as a matter of products liability, negligence or otherwise, or from any use or operation of any methods, products, instructions or ideas contained in the material herein. Because of rapid advances in the medical sciences, in particular, independent verification of diagnoses and drug dosages should be made

British Library Cataloguing in Publication Data
A catalogue record for this book is available from the British Library

Library of Congress Cataloguing in Publication Data
A catalogue record for this book is available from the Library of Congress

ISBN-13: 978-0-240-51999-9
ISBN-10: 0-2405-1999-X

For information on all Focal Press publications
visit our website at www.focalpress.com

Typeset by Charon Tec Ltd, Chennai, India
www.charontec.com
Printed and bound in Canada

05 06 07 08 09 10 10 9 8 7 6 5 4 3 2 1

Working together to grow
libraries in developing countries

www.elsevier.com | www.bookaid.org | www.sabre.org

ELSEVIER   BOOK AID International   Sabre Foundation

Contents

Contents

Contents

Contents

Contents

contents

Mike Collins is a musician and recording engineer/producer who has worked with all the major audio and music software applications on professional music recording, TV and film scoring sessions since 1988.

During that time, Mike has regularly reviewed the latest music and audio software and hardware for magazines including Macworld, MacUser, Personal Computer World, Sound On Sound, AudioMedia, Studio Sound, Electronic Musician, Future Music, and others.

Since 1995, Mike has also been offering personal tuition, consultancy, and technical support, along with lectures and seminars on music technology and music production, typically at Universities and Colleges offering Music Technology courses.

Since the beginning of 2001, Mike has been writing for Focal Press and his first book, entitled 'Pro Tools 5.1 for Music Production' has been available since December 2001. A second book, 'A Professional Guide to Audio Plug-ins and Virtual Instruments' has been available since May 2003. 'Choosing & Using Audio & Music Software', Mike's third title for Focal Press was published in 2004 along with the 2nd edition of the first book, re-titled 'Pro Tools for Music Production'.

Currently, Mike is a Contributing Editor to Macworld magazine in the UK – specializing in music and audio products. Mike also writes industry news and technical reports for Pro Sound News Europe.

Contact details

The Author may be contacted via email at 100271.2175@compuserve.com or by phone at +44 (0)20 8888 5318.

The Author's website can be found at www.mikecollinsmusic.com

Acknowledgements

Thanks to all at Focal Press for the teamwork that produced this book.

Thanks to Digidesign for including me in their beta test program for Pro Tools 7 and for supplying images of their hardware for inclusion in the book.

Thanks to Tom Dambly, Certified Pro Tools Instructor, for reading the proofs to check technical details and for supplying additional information.

Thanks to all the inspiring musicians I have been writing and recording with throughout 2005, in-between writing this book, especially Jim Mullen, Blair Cunningham, Winston Blissett, Damon Butcher, Lyn Dobson, John McKenzie, and Graham Dean.

I would also like to thank my parents, Luke and Patricia Collins, my brother, Anthony Collins, my girlfriend, Athanassia Duma, and my friends Barry Stoller and Keith O'Connell for all the support they have given me – without which this book would never have been completed.

What's in a name?

'Pro Tools' – the name says a lot! I reckon this is a tribute to the power of positive thinking: say what you want to become and you will! Pro Tools was 'born' around 1992 as an aspirational 4-channel recording and mixing system that was more or less a digital version of the popular 4-channel cassette-based Portastudio systems that songwriters used around that time to make their demos. Absolutely nobody considered Pro Tools to be a professional system at the time: it was rightly regarded as nothing more than a toy recording-gadget for computer hobbyists to check out.

But look at Pro Tools today: it has become, without question, the leading system for professional audio recording around the world! The once-mighty Studer multi-track analogue and digital recorders are no longer manufactured, as demand for these machines has plummeted, and even the mighty Neve and SSL mixing consoles are now under threat from Digidesign's ICON mixing worksurfaces.

The truly amazing thing is that almost anyone can afford to buy an entry-level Pro Tools system today – which was never true of a Neve/Studer or any similar professional analogue system. Using Pro Tools LE – the home studio version of the Pro Tools HD software that is used every day in leading professional studios around the world – with hardware options priced to suit most people's pockets – even a kid can get started in a bedroom. And the almost identical Pro Tools M-Powered software works with an even wider range of affordable hardware interfaces.

Who should use it?

Anyone who wants to record, edit, and mix audio and MIDI data will find Pro Tools suitable for these purposes. Recording engineers, music producers, musicians, composers, arrangers, people working on sound in theatre or in music venues – almost anyone who needs to work with recorded audio and MIDI in any situation.

There is plenty of competition around, including Apple's Logic software, MOTU's Digital Performer software, Yamaha/Steinberg's Cubase SX, and

Nuendo software to name just a few. One advantage Pro Tools has over these competitors is that it is unquestionably easier to learn how to use. This is due to the excellence of its design, which manages to fit most features into just two main windows – the Edit and Mix windows – while its competitors confuse their users with far too many windows to fit comfortably onto even the largest computer screen – or even onto a pair of screens.

Students and aspiring recording engineers and producers will particularly benefit from learning how to use the Pro Tools LE or M-Powered software because all the software features in these versions are included in the Pro Tools HD software used in the professional studios – and the user interface is virtually identical. As a result, everything you learn using Pro Tools LE or M-Powered is immediately transferable to the professional Pro Tools HD systems.

Affordable Pro Tools Systems

The Mbox is the most affordable Digidesign branded system, priced at under $500. Avid, Digidesign's parent company, bought M-Audio in 2005. Almost immediately Digidesign released a version of Pro Tools LE called Pro Tools M-Powered that works with M-Audio's expanding range of versatile and affordable audio interfaces. For example, the M-Audio Ozonic is a MIDI keyboard that also incorporates both MIDI and audio interfaces and hooks up to the computer via FireWire. Ideal for laptop users, systems such as Ozonic with Pro Tools M-Powered software are certain to encourage even wider use of Pro Tools systems.

More ambitious setups need more inputs and outputs and possibly a mixing control surface – a box with faders, knobs, and switches that hooks up to the computer to control Pro Tools LE. Digidesign offers the Digi 002 Rack – which has 8 channels of analogue I/O plus digital I/O and monitoring capabilities in a 2U 19″ rack. If you want a control surface, there are two main options: the Digidesign Command | 8 or the Digi 002 – which has the same I/O capabilities as the 002 Rack plus an 8-fader control surface. The Digi 002 looks like a compact 8-channel digital mixer and it can function as a stand-alone 8:2 mixer with two internal effects units plus two external effects sends, EQ, dynamics, other effects, and snapshot automation. The Command | 8 is similar to the Digi 002, but does not include the audio to computer interface – so you need a 002 Rack (or some other suitable interface) as well.

Speakers, Headphones, and Microphones

You also need a suitable loudspeaker monitoring system to work with and at least one set of headphones.

As far as headphones are concerned, the Beyer DT100 or DT150 models are the best choices for regular studio use for the important reason that any of the parts that wear out can be easily replaced. This is not the case with most other headphones – especially those sold for use with consumer audio systems which generally have flimsy cables and earpieces that wear out very quickly with no replacement parts available from the manufacturers.

There are lots of loudspeaker systems available to choose from. Those with built-in amplifiers are usually the most convenient to work with. M-Audio, for example, sells a range of affordable monitors to suit smaller budgets.

But if you really want to do justice to your system, you should go for something like the Mackie HR824 nearfield monitors, which have built-in power amplifiers. The high-frequency amplifier is rated at 100 watts with 150-watt peaks while the low-frequency amp puts out 150 watts with 200-watt peaks – so plenty of headroom to avoid distortion. The 1″ tweeter with its exponential waveguide provides wide dispersion of high frequencies to above 20 kHz while the 8.75″ woofer delivers flat frequency response down to 42 Hz. So these speakers provide lots of bottom end and a comfortable high end with a wide mixing sweet spot – just what you need for long sessions.

And if you want something even punchier, my top recommendation has to be the ATC SCM SCM20ASL Pro active monitors. These 'deliver' whether you are recording, mixing, or mastering. ATC monitors are widely used in top mastering, orchestral and music studios around the world. These models use a 6″ combination mid/bass drive unit and a 1″ tweeter to deliver performance that you would normally expect only from a large 3-way monitor. Each cabinet has a 250-watt amplifier for bass/mid and 50 watts for high frequencies – and will deliver continuous Sound Pressure Levels of up to 108 dB. Yet even at the highest-volume levels, the amp and driver system maintains ultra-low distortion, solid stereo imaging, and a high degree of linearity. So what you hear is what you got!

If you have to choose just one versatile and affordable microphone for vocals and instruments, this has to be the Shure SM58 or SM57. If you have a little more to spend, I recommend Beyer M130, M160, and M260 ribbon microphones. The M130 and M160 can be arranged in a stereo M/S configuration that sounds great for recording pianos, drum kits, and almost any acoustic or electric instruments. The M260 is ideal for recording acoustic guitars. And if your budget will stretch, a Neumann M149 or M147, or an AKG C12VR or C414 studio condenser microphone will let you capture audio at the highest quality.

Home Studio Setups

A very basic home setup will just be in one room or part of a room. A separate vocal booth is always a good idea – whether you are recording voiceovers or vocals or one or two instruments, singers, or musicians. A separate recording room or studio is better still.

Don't forget that human beings have to breathe fresh air to function properly, so a suitable ventilation or air conditioning system should be regarded as a necessity – not a luxury. Seating and benching should be ergonomically designed so that whoever is sitting and working in the room is not having their attention constantly distracted because they can't sit comfortably or reach the equipment properly. You should also pay attention to the lighting in the room, making sure that this can be bright enough when it needs to be and dim enough when a more intimate atmosphere is required.

If you don't want to disturb your neighbours or have them disturb you while you are recording, you are going to have to consider soundproofing. If an airplane flies past or a truck rumbles by, or someone in the building slams a door while you are recording into microphones, these sounds will be there in your recordings and can be impossible to remove completely afterwards without adversely affecting the sounds you want. You can buy acoustic tiles, baffles, panels, and other devices to add to a room to cut down any excessive reverberation or reflections of sound, but these won't stop sounds coming in from outside or prevent all your sounds from travelling outside. You can get some useful isolation by constructing inner floors, ceilings and walls, and filling the area between outer and inner with layers of acoustically absorbent materials such as rockwool. Nevertheless, the only way to do this truly effectively is to use very expensive sound isolation techniques such as suspending a room within a room on vibration isolation mountings to prevent sound entering or leaving structurally and/or lining the room with lead to damp any vibrations before they can be retransmitted from the walls.

Laptop Musicians

With a laptop you can record on top of a mountain or at the bottom of the Grand Canyon or at any point in between that you can reach.

Laptops are very popular with musicians, especially now that there are so many virtual instruments available, ranging from classic instruments such as the Hammond Organ, Steinway Piano, Fender Rhodes, Moog, ARP, Yamaha, and Korg synthesizers, to the simulations of most popular guitar amplifiers and effects units that can be found in Native Instruments Guitar Rig and IK Multimedia's AmpliTube.

If you are on the move with your laptop and you want to set it up to make music in a hotel room or in a field or in someone else's studio, you will probably want a small keyboard so that you can enter MIDI data with a built-in MIDI interface or USB connection to connect it to your computer, and maybe some knobs and sliders to control software parameters on-screen. And if you want to carry your studio on your back in style, M-Audio, for example, makes a padded nylon Studio Pack. This conveniently carries an M-Audio Ozone or Oxygen 8 keyboard along with a laptop and accessories.

Mac or PC?

Pro Tools was originally developed on the Mac and, typically, the feature set on the Mac has always been a little ahead of the feature set on Windows.

Nevertheless, Digidesign is committed to achieving parity between the software for the two platforms wherever possible, and it has to be said that there is little, if any, practical difference between the two implementations these days.

I personally favor the Mac platform because I like the Mac's OS X operating system better than Windows and I have always found the Mac to be easier to troubleshoot than my Windows computers. I also use various software applications that are only available on the Mac platform, such as Digital Performer.

However, there are good software applications for Windows that are not available on the Mac, such as Sonar and GigaStudio, which is an argument in favor of owning two computers for music so that you get the best of both worlds.

Perhaps the ideal would be to buy a Mac that uses an Intel processor and allows you to run Windows as well as OSX. It is rumoured that the next generation of Apple computers will do just this. But they said that about the PowerPC platform, and it never happened. So we will just have to wait and see. . .

Pro Tools LE Software

Pro Tools LE software serves as the interface between you, your computer, and your Pro Tools LE hardware. It lets you record and play back up to 32 tracks of audio at once and allows you to import and export audio using various popular file formats. It is also a full-featured MIDI sequencer that can record, edit, and mix up to 256 MIDI tracks alongside your audio. MIDI events and sequences can be manipulated as easily as audio in Pro Tools software, in most cases sharing the same editing and mixing tools.

You can adjust every aspect of the audio and MIDI tracks using a single screen – the Edit window. Here you can fix the guitarist's 'bum' notes, get rid of any pops or clicks, instantly change the tempo of your audio files and loops, rearrange your songs, stretch or squeeze your audio to fit to picture, and so forth.

And when it's time to mix, you can use the Mix window to automate every move you make with the audio and MIDI tracks, and to automate every effect that you have applied using the plug-ins – giving you near total recall of your mixes. You can also send and return audio to and from outboard gear via the audio interface.

Pro Tools M-Powered Software

Pro Tools M-Powered software is a special version of Pro Tools LE designed to work with various affordably priced M-Audio interfaces.

Pro Tools M-Powered software is virtually identical to Pro Tools LE. The main differences are that M-Powered does not include the Ignition Pack of plug-ins and does not support the Control | 24 control surface, the DV Toolkit option, or the DigiTranslator option that allows you to import and export files compatible with other digital audio and video software.

You can open sessions created in Pro Tools M-Powered software with any Mac OS X-based or Windows XP-based Pro Tools system, allowing you to easily collaborate with other Pro Tools users. So you can take your projects to a Pro Tools HD-equipped recording studio for additional recording or mixing – or take professional studio projects back home or on the road.

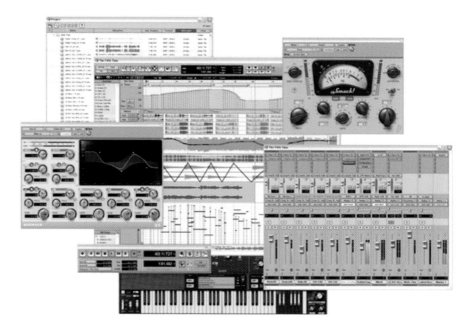

Fig. 1 – Pro Tools M-Powered software.

Transfers

Pro Tools lets you easily transfer sessions between your home setup and a professional studio or onstage rigs – encouraging collaborations with other Pro Tools users.

All Pro Tools sessions can be transferred between any types of Mac-based or Windows-based systems – whether you are using Pro Tools LE, M-Powered, or HD systems.

Included Software

Every Pro Tools system includes the standard DigiRack plug-ins along with seven Bomb Factory plug-ins: the BF76 Compressor, the Essential Clip Remover, the Essential Correlation Meter, the Essential Meter Bridge, the Essential Noise Meter, the Essential Tuner, and the Funk Logic Mastererizer.

Current Pro Tools systems also ship with the Pro Tools Ignition Pack. This includes Propellerhead Reason Adapted, Ableton Live Lite Digidesign Edition,

FXpansion BFD Lite, Celemony Melodyne 'uno essential', and T-RackS EQ, AmpliTube LE, and SampleTank 2 SE from IK Multimedia.

Reason Adapted gives you a virtual rack of MIDI-based synthesizers, samplers, drum machines, effects to work with. Audio from Reason Adapted can be streamed directly into the Pro Tools mixer via ReWire for processing and mixing within Pro Tools.

Live Lite Digidesign Edition lets you compose, record, remix, improvise, and edit musical ideas using its intuitive sampler and sequencing features. Live's 'elastic audio' sequencing modifies the tempo and pitch of loops in real time, allowing you to drop multiple loops into a song regardless of pitch or tempo. Individual outputs from Live Lite can be streamed directly into the Pro Tools mixer through ReWire.

BFD Lite provides you with three meticulously recorded drum kits along with additional individual samples – all with multiple velocity layers – to use with your Pro Tools system.

SampleTank 2 SE is an easy-to-use sample playback module that provides quick access to a wide range of sounds. This plug-in has an excellent library of samples and instruments and features powerful playback engines with high-quality built-in effects.

AmpliTube LE is a flexible guitar amp-, cabinet-, and effects-modeling plug-in with a variety of amp and cabinet models, classic stomp boxes, and other effects to choose from. Though AmpliTube LE sounds great on guitar, it can be used on almost anything – from vocals to drum tracks.

T-RackS EQ is a warm sounding tube-modeled parametric EQ that provides a useful alternative to the standard DigiRack EQs for mixing and mastering.

Melodyne 'uno essential' lets you correct the tuning in mono audio tracks, edit the timing of melodic lines, or completely rearrange parts through its intuitive user interface.

These basic versions give you a 'taste' of what you get with the full versions, so if you find these useful, you should consider upgrading to the full versions, which have a lot more to offer.

The Ignition Pack also includes the Pro Tools Method One instructional DVD, a Bunker 8 REX File CD, 1-year membership to Broadjam.com and a limited-time free trade magazine subscription.

The Pro Tools Method One instructional DVD helps you learn Pro Tools system essentials, from setting up sessions and recording audio to editing MIDI, working with loops and plug-ins, automating mixes, and more.

The Bunker 8 REX File CD is a useful collection of REX files for your Pro Tools system.

Even more software

There are also various extra bundles of software that are either included for free with particular Pro Tools systems, or that can be purchased at attractive prices.

For example, Digi 002 systems include the following software bundle at no additional charge: Bomb Factory BF-3A classic compressor; Moogerfooger Ring Modulator and Analog Delay based on Bob Moog's classic designs; Cosmonaut Voice, which offers a telephone voice and interplanetary effects; JOEMEEK SC2 Photo Optical Compressor and VC5 Meequalizer; Tel-Ray Variable Delay, SansAmp PSA-1; Voce Spin rotary speaker simulator and Voce Chorus/Vibrato.

For just under $500, Digidesign offers the Music Production Toolkit, an impressive selection of tools that can help easily expand the creative power of a Pro Tools LE or Pro Tools M-Powered system. With the Music Production Toolkit, users can take advantage of an impressive collection of plug-ins and enhanced editing tools, increased track count, and the ability to easily export mixes as MP3 files.

The Music Production Toolkit includes a host of powerful plug-ins ideal for creating and producing music, including: Hybrid, a versatile high-definition synthesizer from Digidesign's new Advanced Instrument Research group; TL Space Native Edition, a pristine convolution reverb; Smack! LE, a professional compressor/limiter; SoundReplacer, a time-saving sound replacement tool; and Digidesign Intelligent Noise Reduction (DINR) LE, an easy-to-use noise reduction plug-in.

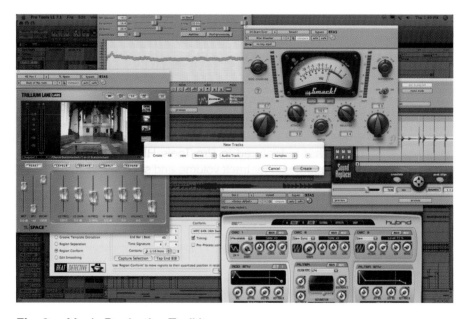

Fig. 2 – Music Production Toolkit.

With the Music Production Toolkit, users can also expand their Pro Tools sessions up to 48 mono or 48 stereo tracks at up to 96 kHz for more complex mixes. In addition, the Music Production Toolkit features a multi-track version of Beat Detective, which allows users to perform automatic groove analysis and correction across multiple audio and MIDI tracks at the same time. This multi-track capability was previously available only with Pro Tools HD software. The Toolkit also provides the Pro Tools MP3 Option for exporting mixes as MP3 files. Finally, the Music Production Toolkit includes an upgrade to the latest Pro Tools software (currently version 7.1) from any previous version of Pro Tools LE or Pro Tools M-Powered software.

What you don't get in Pro Tools LE and M-Powered

So what do you miss out on with Pro Tools LE and M-Powered compared with Pro Tools HD software? Two of the main differences are that you are restricted to 32 tracks and there are no surround features. Beyond that, there is no Continuous Scrolling option, no Momentary Solo Latch, no Shuttle keypad mode, no Collection Mode in Beat Detective, no Universe window, and no Delay Compensation. Some of these are not too essential, but I particularly miss Pro Tools HD's Edit menu 'Play Timeline Selection', 'Change Timeline to Match Edit Selection', and 'Change Edit to Match Timeline Selection' commands.

With Pro Tools HD, you can choose to have Pro Tools automatically apply real-time fade-ins and fade-outs to all region boundaries in the session, which saves you the trouble of editing to zero-crossings or creating numerous rendered fades in order to eliminate clicks or pops during playback. Pro Tools LE and M-Powered do not have this Auto Region Fade feature. The Trim tool does not have Pro Tools HD's Scrub Trim feature that lets you scrub along a track to listen for a trim point, then automatically trim to where you release the mouse button. And the Track Punch feature that makes Pro Tools HD more useful as a digital dubber for film re-recording and mixing, for loading dailies and recording Foley, as well as for over-dubbing and tracking in music sessions, is not available.

There are also some limitations when it comes to mixing. For example, the Trim mode that is featured in Pro Tools HD that lets you modify already written automation data for track volume and send levels in real time is not supported. Also, there is no Snapshot Automation and no 'Copy to Send' command. 'Copy to Send' lets you copy the entire automation playlist for the selected control to the corresponding playlist for the send. This is very useful if you want a track's Send automation to mirror the automation in the track itself so that the effect level follows the levels in the main mix.

Another area where features are more limited is synchronization. There is support for the Generic MTC Reader in the Peripherals dialog, so you can use the MOTU MTP AV or Digital Timepiece, for example, but there is no support for any of Digidesign's sync peripherals such as the SYNC I/O. And although there is support for MIDI Machine Control (MMC), there is no support for 9-pin Serial or 9-pin Remote Machine Control. The good news is that you can lock to all MTC formats and use MMC. However, there is no Timecode or Feet and Frames display, no Pull up/Pull down, no Movie Sync Offset, and no Auto Spot mode, and although it supports MIDI controllers such as the HUI, there is no Pro Control support.

The bottom line? These missing features are particularly important for video post-production and film work and most can be added using the DV Toolkit option (see Appendix 2). If you are working mostly in music production you can live without these features most of the time, although you can always add the Music Production Toolkit option (described in the previous section) to get more tracks and the extra Beat Detective features.

What's in the book

Chapter 1 presents descriptions and notes about Digidesign's Pro Tools hardware for LE systems along with overviews describing M-Audio's hardware for M-Powered systems.

Chapter 2 offers a more detailed description and notes about the Digi 002 control surface and interface along with an overview of the Command | 8 control surface which has similar features.

These first two chapters will be very helpful to anyone who is planning to buy or upgrade their Pro Tools LE or M-Powered system and wants to know what the hardware options are.

In Chapter 3 you will find descriptions of the Pro Tools LE software, explaining various features of the Mix, Edit, and Transport windows, along with overviews of all the menus and of the file browsers.

This chapter is intended to inform the reader about the capabilities of the software and to provide useful tips and notes about the user interface. It will be useful for readers who want to compare Pro Tools with similar software, and as a 'primer' for anyone 'on the learning curve' who will find the highlighted notes and tips draw attention to important details that they may have missed when reading the manuals.

note   ▷   Descriptions of Pro Tools LE software apply equally to Pro Tools M-Powered software – which is virtually identical.

Chapter 4 explains how to restore the default preferences so that anyone who wishes to follow the step-by-step examples featured in this and subsequent chapters will see similar looking screens. Lots of basic information about opening new sessions, saving sessions, allocating hard disk space, setting up tracks, and importing audio and video files is presented here.

Chapter 5 is all about tempos and grooves. If you are recording musicians, it is important to make sure that they are playing in time with a click so that you can edit the performances afterwards. If you are working with audio that was not recorded to a click, you will need to adjust the tempo of the Pro Tools session to that of the recorded or imported audio (or vice versa) to allow editing. Beat Detective is also explained with examples.

One of the big steps forward from Pro Tools 6 to Pro Tools 7 is the enhanced MIDI sequencing capabilities. Tied in with this is the rise and rise of virtual instruments which seem to be getting better, more numerous, and more popular each day that passes. Chapter 6 explains how to get started using the virtual instruments supplied with Pro Tools systems and how to record these to audio tracks ready for mixing in Pro Tools.

Editing MIDI is covered in Chapter 7, which explains the editing features that are available then provides step-by-step examples to give you a 'jump-start'.

Chapter 8 shows you how to prepare to record audio before guiding you step-by-step through your first audio recordings and first edits.

Pro Tools gives you lots of tools to edit audio with, so Chapter 9 is packed full of explanations, notes, and tips about how to use these editing features.

Chapter 10 aims to get you up to speed quickly with the technical features that you will use during your mixing sessions. Mixdown to stereo is also covered in some detail.

At the end of the book you will find a couple of useful appendices. Appendix 1 talks about backups and transfers while Appendix 2 explains the features of the DV Toolkit. To finish up, the (Further Information) More Info section has a list of recommended books, magazines, and websites that you can check out if (when!) you are hungry for more information.

Mike Collins © 2006

1

Hardware Options for Pro Tools LE and M-Powered Systems

Pro Tools LE software requires you to connect a suitable Digidesign hardware interface to your computer before it will run. Digidesign hardware for LE systems includes the Digi 001, the Mbox, the Mbox 2, the Digi 002, and the Digi 002 Rack. The Command | 8 hardware controller also works with LE systems.

The Mboxes are very affordable and ideal for use with laptops but only offer two channels of analogue and two channels of digital input and output. The Digi 002 Rack is more expensive, but supports multi-channel operation. The Digi 002, covered in the next chapter, incorporates a hardware control surface along with the same interfacing features as the Digi 002 Rack. The Command | 8, also covered in the next chapter, is similar to the Digi 002 Rack, incorporating a MIDI interface and basic audio monitoring, but without the multi-channel audio input and output interface capabilities.

Let's take a look at the discontinued Digi 001 first, as this was the first serious multi-port interface for Pro Tools LE systems and there are still many of these in use today.

Digi 001

The Digi 001 system includes a PCI card that you install in your computer; an I/O interface with a range of input and output connectors; and the Pro Tools LE software. The PCI card has a connector to link to the Digi 001 interface and also has a pair of ADAT optical connectors for multi-channel digital input and output (I/O). Each optical connector can carry 8 channels of 24-bit digital audio or can be switched using the software for use as an additional 2-channel S/PDIF interface.

The I/O interface provides 8 analogue audio inputs and outputs with 24-bit converters, S/PDIF digital audio input and output, a stereo headphone jack socket, a footswitch jack socket, and a pair of MIDI In and Out sockets. The total number of audio inputs and outputs is 18 – counting the 8 ADAT digital channels, the 2 S/PDIF digital channels, and the 8 analogue channels.

Fig. 1.1 – Digi 001: no longer manufactured.

The MIDI in and out connectors let you use the Digi 001 as a basic 16-channel MIDI interface for Mac or PC. The headphone jack socket lets you connect a pair of stereo headphones to monitor whatever audio you have routed to analogue outputs 1 and 2 using the Pro Tools LE software. The footswitch jack lets you connect a footswitch to control the QuickPunch and MIDI punch-in and punch-out recording features.

The 001's microphone preamps will accept a wide range of microphone types, including high-quality 'condenser' models that require 'phantom' power. Dynamic microphones (such as the popular Shure SM58) don't need power, but professional studio models (such as the AKG C414) need 48 volts which the preamplifier can send via the microphone cables into the microphone, rather than via separate power cables – hence the name 'phantom' power.

Fig. 1.2 – Digi 001 front panel.

The pair of analogue audio inputs provided on the front panel of the interface can accept either Mic-level or Line-level signals. The input sockets are combined XLR and 1/4″ jack types, so you can plug either of these commonly used connectors into the Digi 001. The 'pad' switches provided for each input reduce the input sensitivity by 26 dB when you want to use line-level signals instead of microphones. The phantom power switches are labelled '48 V' which is the voltage supplied to power the microphones. A pair of Input Gain controls is also provided. Gain controls for the line inputs, 3–8 on the back of the Digi 001 interface, are provided in software so you can use these with mixers, preamps, keyboards, or other line-level sources.

Fig. 1.3 – Digi 001 rear panel.

A pair of Monitor Outputs is provided on the back panel. These carry the same audio signals that are routed to analogue outputs 1 and 2. The difference is that they are intended to let you listen to your main mix by hooking up a pair of powered speakers or a stereo power amplifier and speakers. A Monitor Volume control is provided on the front panel of the Digi 001 interface to control the listening level.

The Main Analogue outputs, 1 and 2, can be used to connect to a tape recorder when you are mixing down, although they may be connected to an external mixer. These outputs are balanced, +4 dBu line level. Analog Outputs 3–8 are unbalanced, −10 dBu line level, and can be used as sends to outboard gear or as outputs to an external mixer.

Another option for mixdown is to connect the S/PDIF outputs to a DAT or other digital recorder. By default, whatever is routed to outputs 1 and 2 is also sent to the S/PDIF outputs. You can disable this S/PDIF 'mirroring' (as it is referred to in the manual) so that you can use the S/PDIF connections to hook up an external digital effects unit, for example.

Low Latency Monitoring on the Digi 001

The Digi 001 supports Pro Tools LE's special Low Latency Monitoring feature. This lets you record and monitor the 8 analogue and 8 ADAT digital inputs directly via outputs 1 and 2 with very low latency. When you select Low Latency Monitoring from Pro Tools LE's Options menu, the audio entering the hardware interface is not passed through the host computer's processor – it is routed directly to the main outputs instead.

There is a slight amount of latency involved here due to the A/D and D/A conversions and the internal routing within the hardware. From the information I was able to gather, the total delay is roughly 164 samples. This may not sound that low, but it's a lot better than the 1024 samples that would have been in there had the Low Latency feature not been implemented.

3

There are some limitations with this method – any plug-ins and sends assigned to record-enabled tracks will be bypassed, for example – but at least you can get around the problem reasonably well.

One particular situation always requires a workaround. If you are using a plug-in to create a metronome click, you are going to lose your click in Low Latency mode.

One solution is to use an external MIDI device such as a drum machine to play the click. Another, partial, solution is to bus the output of the track containing the plug-in to an audio track and record this to your hard disk first. Then you can dispense with the plug-in track and simply replay the audio click track.

The problem with this, of course, is that you lose the ability to hear the click during count-off bars. If your session starts at Bar 1, Beat 1, for example, this can make it practically impossible to cue a musician to overdub starting at Bar 1, Beat 1.

A solution here is to make sure that you leave one or more bars empty at the start of your session so that you can record the click as audio into these 'startup' bars and use this to cue the musicians.

Monitoring latency and MIDI

To monitor a MIDI device through the audio inputs on the Digi 001, you need to route each input to a track and record-enable that track before you will hear any output. This is another reason why it is useful to use a separate external mixer with the Digi 001 – so you can always hear your synths, drum machines, and samplers without setting up routings in the Pro Tools software.

Also, when you are monitoring the audio coming into the Digi 001 from an external synthesizer, what you hear will have an audio delay equivalent to the number of samples specified in your Hardware Buffer settings – the latency delay. This delay will be very apparent if you have existing audio tracks, as these will be heard first and the MIDI devices will be heard a little later. If you don't have an external mixer you will have to accept this latency while recording MIDI.

But there is a way around this for playback – use the Global MIDI Playback Offset feature in Pro Tools LE's MIDI Preferences to trigger your MIDI data early to compensate for the latency. This offset is made in such a way that it just affects the playback – not the way the MIDI data is displayed in the Edit window. You are given the choice of offsetting the MIDI tracks either globally (all by the same amount) or individually. To compensate for audio monitoring latency you will need to enter a negative offset that causes the MIDI data to be played back earlier by a number of samples equivalent to the latency in samples. The best way to work out which latency value to use is to record the audio from your MIDI device into Pro Tools, then simply look at the exact position in samples where the audio starts compared to where the MIDI note is placed. This way you can read off the delay between these exactly in samples.

tip ▷

You can also set up Individual MIDI Track Offsets in Pro Tools LE. This can be useful when you want to compensate for the time it takes for a particular synth or sampler to respond to an incoming MIDI message. This can amount to several milliseconds and this can be enough to make supposedly simultaneous percussive instruments sound like they are 'flamming'. This is typically the case with a snare drum sample played from Pro Tools as audio that you want to combine with another snare sample played from an external MIDI sampler. The solution here is to offset the individual MIDI track in Pro Tools to compensate for the delay in the MIDI sampler.

Digi 002 Rack

Priced around $1200, the Digi 002 Rack is, essentially, the Digi 002 without the control surface. What you get is a 2U 19″ rackmountable unit that connects to the computer via FireWire. It is just about light enough in weight to carry with you in a bag, and it is perfect for small studio setups – supporting a wide range of analogue and digital audio, and MIDI inputs and outputs. It also has dedicated monitor and headphone outputs.

Fig. 1.4 – The Digi 002 Rack: the successor to the Digi 001.

The Digi 002 Rack will record audio with 24-bit clarity and up to 96 kHz sample rate support

Front panel

Fig. 1.5 – Digi 002 Rack front panel.

At the left of the front panel there are rotary gain controls for each of the four microphone preamplifiers. Above these there are four pairs of buttons. The first of each pair switches the input between microphone and line level. The second button switches in a high-pass filter to remove lower frequencies that may contain rumble or other unwanted sounds. Four LEDs to the right of the gain controls indicate the sampling rate as: 44.1, 48, 88.2, or 96 kHz. Three more LEDs to the right of these indicate MIDI input and output activity. Above these there are two small button switches. The first routes the Alternative Source to input pair 7–8 while the second routes the Alternative Source to the monitor outputs. An associated rotary control knob lets you adjust the output level to the monitors and, adjacent to this, a button switch is provided to let you mute the monitors. Finally, at the right of the panel, there is a 1/4″ jack socket, providing a headphone output, with an associated volume control knob.

Rear panel

Fig. 1.6 – Digi 002 Rack rear panel.

Looking at the rear panel you will find the inputs at the right and the outputs at the left. The first four inputs accept either microphones via XLR connectors or line- or instrument-level inputs via 1/4″ jack connectors. Two switches are provided for 48-volt phantom power – the first switches this on for inputs 1 and 2, while the second switches it on for inputs 3 and 4.

This is something of a limitation if you only have one microphone that needs phantom power. Fortunately, most dynamic microphones are not usually harmed if phantom power is applied, even they do not need this. Unfortunately, ribbon microphones can be damaged if you apply phantom power – so take care!

note ▷ So-called 'phantom' power is used to supply the 48 volts typically used to power capacitor microphones directly via the microphone audio cables. Examples include the large-diaphragm studio types like the AKG C414 or Neumann U87 models.

Inputs 5, 6, 7, and 8 just have 1/4″ jack connectors, but individual switches are provided for +4 dBu or −10 dBV operation. An Alternate Source Input for direct monitoring of −10 dBV audio equipment, such as tape players or CD players, is provided via a pair of RCA/phono sockets for left and right channels.

To the left of these, a pair of 1/4″ jacks sockets carries the left and right Monitor outputs. You connect these to your monitor amplifier and speakers, or directly to powered monitors, and you can control the level of this output using the dedicated volume control on the front panel. This monitor output carries whatever signals are being routed to Main output pair 1–2.

The main left and right outputs, output pair 1–2, may be connected to a tape or DAT or CD recorder when recording final mixes to stereo, or may be used along with the other three output pairs, 3–4, 5–6, and 7–8, to connect to any other equipment.

The Monitor outputs and the individual outputs are fixed at +4 dBu to suit connection to professional equipment. Signals routed to output pair 1–2 are also mirrored on a pair of RCA/phono sockets that provides an alternative −10 dBV output to, say, a cassette or Mini-disc recorder.

So, in total, the Digi 002 Rack has 8 analogue input channels and 8 analogue outputs, and the A/D and D/A converters offer 24-bit/96 kHz operation – which provides significantly higher audio quality than the Digi 001 or Mbox, for example.

One MIDI input socket and two MIDI output sockets are provided to support 16 channels of input and 32 channels of output. To the left of these there is a pair of optical connectors that can provide either 8 channels of ADAT optical I/O (up to 48 kHz) or 2 channels of S/PDIF I/O (up to 96 kHz).

There are also two RCA/phono sockets for standard S/PDIF I/O and a pair of IEEE 1394 'FireWire' ports – one to connect to your computer and one to act as a 'thru' port to another FireWire device, such as an external hard drive.

A 1/4″ jack socket to the left of the FireWire ports, labelled Footswitch, is provided for QuickPunch control – use this to punch in and out while recording.

Finally, the power connector will accept AC supplies between 100 and 240 volts running at either 50 or 60 Hz – so you can connect the Digi 002 Rack to the main electricity supplies in most countries around the world without problems.

The Mbox

Mbox is a low-cost system for recording, editing, and mixing up to 32 tracks of audio that includes an audio interface along with Pro Tools LE software. The interface connects to your computer via USB – ideal for use with laptops.

The Mbox requires a free USB port on the computer to connect to – it will not work with a USB hub – which could make things awkward with some setups. However, the Mbox takes its power from the USB port, so no power adapter is required – a distinct advantage for portable setups.

Compact in size, Mbox measures just over six inches high, three and a half inches wide, and seven inches deep. It has two analogue audio inputs into which you can feed a microphone, a line-level signal, or a direct instrument signal via the XLR/1/4″ jack combination connector. The high-impedance line input has a low-noise setting for use with electric instruments such as guitar, bass, or keyboard. MBox also has 2 channels of S/PDIF digital I/O.

The A/D and D/A converters and the S/PDIF I/O are all 24-bit, working at either 44.1 or 48 kHz sampling rates, and the microphone preamplifiers were designed by Focusrite to provide the best quality possible at this price.

The front panel

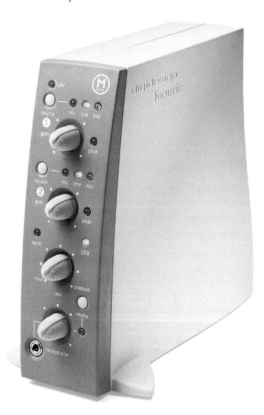

Fig. 1.7 – The original Mbox: now discontinued.

Looking at the front panel, there are two knobs with associated switches and LED indicators to let you select between the mic, line and instrument inputs, and control the input levels of these. Below these, third from the top, you will find a knob labelled 'Mix' that lets you monitor whatever is plugged into the

Mbox's inputs, or the playback of any audio tracks in Pro Tools, or a blend or mix of these, as described later. Finally, at the bottom of the front panel there is a 1/8″ 'mini' headphone jack with an associated volume knob and mono button.

The back panel

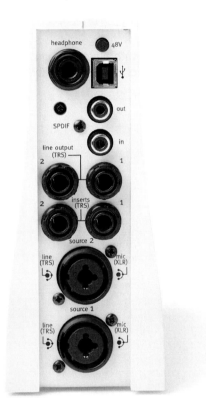

Fig. 1.8 – Mbox back panel.

At the top-left of the back panel there is a standard 1/4″ headphone jack. To the right of this there is a USB port with a 48-volt phantom power switch above it. Underneath the USB port there is a pair of RCA/phono connectors for S/PDIF input and output. The line outputs are underneath again, in the form of a pair of 1/4″ TRS jacks that can provide either balanced or unbalanced analogue output. A pair of analogue inserts is also provided via 1/4″ TRS jacks to allow you to hook up outboard processors while recording to disk. The final two connectors are dual-purpose XLR/1/4″ jack combinations that will accept either a typical balanced microphone cable XLR connector or a typical musical instrument cable 1/4″ jack connector.

Zero-latency monitoring

One of the neatest features is the Mix control. This lets you blend the sound of anything plugged into the Mbox with the playback from Pro Tools LE. This is ideal, especially when overdubbing, to avoid the latency delay that normally appears when you monitor inputs through the software (via A/D, the computer CPU, then D/A) and back out to the speakers – rather than monitoring straight from the source to the speakers.

Just mute the track to which you are recording in the Pro Tools LE software mixer so that you don't hear its delayed sound being monitored through the system. Then adjust the Mix control to feed the incoming signal from the instrument you are recording directly into the stereo mix. Et voila! There you have it – zero-latency monitoring.

note ▷ When you are using the Mbox hardware with Pro Tools LE software, there is no Low Latency Monitoring option in the Operations menu. This is not needed, as the Mix control effectively does the same thing.

Exploring the Mbox Mix control

Try this:

Step 1. Plug a microphone into one of the Mbox inputs.

Step 2. Route this to an audio track in the Pro Tools LE mixer.

Step 3. Record-enable the audio track so that the audio coming in from the Mbox will be routed via the Pro Tools mixer back out to the Mbox.

Step 4. Now hit the Record and then the Play buttons (or hit Command-Spacebar) to start recording and count out loud to ten (or just say anything for 10 seconds or so) into the microphone.

Step 5. Then hit the Spacebar to stop recording and de-select the track's Record-Enable button.

Now let's explore how the Mix control works:

Step 6. Turn the Mix control to Input (all the way anti-clockwise, to the left).

Step 7. Speak into the microphone. You will hear the microphone directly through the speakers.

Step 8. Now turn the Mix control to Playback (all the way clockwise, to the right).

Step 9. Speak into the microphone and you will discover that you cannot hear the sound directly through the speakers any more.

Step 10. Now hit the Spacebar to playback your recording. You will hear this directly through the speakers.

Step 11. Now put the Mix control back to Input and the playback from Pro Tools will not be heard any more.

This is because putting the Mix control in the Input position only lets you hear what is plugged into the Mbox's inputs (with no latency delay as it is not being monitored via the computer) and putting it in the Playback position only lets you hear what is being played back from Pro Tools.

note ▷ Beware the trap of forgetting to turn the Mix control to either Input or Playback as necessary or you may end up 'scratching your head' wondering why you are hearing no sound through the speakers at all.

tip ▷ The clever trick is to balance the Mix control somewhere in-between these two extremes so that you hear enough of the Input directly via the Mbox so that you can play or sing into the microphone without being put off by the latency delay while hearing enough of the Playback to keep you happy as well.

Summary

Ideal for laptop users, the Mbox interface is very compact, light, and easy to carry around in a bag with a laptop, a pair of small self-powered monitors, and maybe an M-Audio Oxygen keyboard and a Shure SM58 microphone. This setup is basically a 'studio-in-a-bag' that you can set up just about anywhere you like.

Fig. 1.9 – Laptop with Mbox.

Reasons to buy an Mbox? If you want a portable recording studio for your laptop, get an Mbox. If you want to learn Pro Tools before investing in an expensive professional system, get an Mbox. If you are on a budget, but want to make a start working with the industry-standard digital audio workstation, get an Mbox. If you already have a Pro Tools system at the studio and want to take work home to review and edit, get an Mbox.

Mbox 2

Digidesign brings out new products each year and retires older models from time to time. The original Mbox was discontinued in the summer of 2005 and immediately replaced in Digidesign's product range by the Mbox 2 – allowing Digidesign to continue offering a USB-powered studio-in-a-box for under $500.

The Mbox 2 is slightly larger than the original Mbox and features MIDI in and out ports in addition to similar audio features to the original Mbox. It also has microphone, line, and DI inputs. So you can hook up a MIDI module or controller, then plug a guitar or bass directly into the DI jack (without the need for an extra preamp), plug a keyboard or mixer into the line jack, or plug a pair of microphones into the XLR jacks.

Although Mbox 2 looks like a 2-channel unit, it actually supports 4 channels of input and output if you use both the analogue and digital I/O at the same time. As with the original Mbox, the A/D and D/A converters and the S/PDIF I/O are all 24-bit, working at either 44.1 or 48 kHz sampling rates.

The maximum analogue input level is +21 dBu and maximum analogue output level is an impressive +4 dBV– so Mbox 2 should interface with just about any type of mixer or other audio equipment that you are likely to be using.

The Mbox 2 hardware unit can be positioned horizontally or vertically to fit into any workspace. It comes with two faceplates that can easily be swapped using the included hex wrench. One faceplate features a large handle that makes the unit easily portable. The second faceplate, which offers a rubberized grip, allows Mbox 2 to sit flat on a desktop and reduces the space required when the unit is packed for travelling.

Weighing in at just 1.23 kg, Mbox 2 is compact and easily portable. The dimensions measured with the unit in horizontal (handle-down) configuration are as follows: height: 4.9 cm, with handle: 7.7 cm, with handle cover: 5.6 cm; width: 22.4 cm; depth: 18.4 cm (including the knobs).

Fig. 1.10 – Mbox 2.

Front panel

At the far left there is a 1/4″ headphone jack with an associated knob to control the volume. To the right of this you will see the main Monitor volume control and the Mix control that lets you balance between incoming source audio being routed direct to the monitor outputs and audio from the computer being routed to the monitor outputs.

tip ▶ Use the Mix control for zero-latency monitoring of your source audio while recording. With it somewhere in the middle you can hear both source audio and audio from the computer.

Looking at the right-hand side of the front panel you will see the Input gain controls along with associated indicator LEDs, source selection, and Pad buttons.

In the middle there is a very useful Mono button and a button to switch on the 48-volt 'phantom' power, which is needed if you want to connect condenser microphones directly to Mbox 2's microphone preamplifiers.

Fig. 1.11 – Mbox 2 front panel.

Rear panel

Toward the left-hand side, the rear panel has two groups of analogue inputs, each with an XLR microphone input, a balanced Line input (via 'stereo' 1/4" Tip/Ring/Sleeve jack) and an unbalanced DI input (via 'mono' 1/4" jack) to let you connect musical instruments.

In the middle of the panel there is a pair of 1/4" jacks carrying the left and right unbalanced analogue outputs. Next to these, a pair of RCA/phono jacks provides stereo 24-bit S/PDIF digital input and output. To the right of the panel there is a pair of sockets for MIDI input and output and at the far right you will find the USB connection that carries data between Mbox 2 and the computer.

note ▷   The USB socket also carries power from the computer for Mbox 2 so it needs to be connected directly to the computer or via a powered USB hub.

Fig. 1.12 – Mbox 2 rear panel.

M-Audio M-Powered Interfaces for Pro Tools

M-Audio offers a range of PCI, USB, and FireWire hardware interfaces that work with Pro Tools M-Powered software – providing many more options for users to put together suitable, affordable Pro Tools systems.

FireWire vs USB

There are two types of connectors used to hook up popular Audio and MIDI interfaces to your computer – USB and FireWire (IEEE 1394). FireWire supports much greater bandwidth than USB 1.1 so you can use more channels of digital audio at higher sampling rates and bit depths with your application. USB is OK for less demanding applications with just a couple of channels of input and output. FireWire is better able to handle multi-track operation with lower latency and higher fidelity.

FireWire interfaces

M-Audio offers five FireWire multi-channel audio/MIDI interfaces that work with Pro Tools M-Powered – the Ozonic keyboard and interface, the FireWire 410 and FireWire 1814, FireWire Audiophile, and FireWire Solo.

Ozonic

Ozonic is a 37-note velocity- and pressure-sensitive MIDI keyboard, with plenty of MIDI-assignable knobs, sliders, buttons, joystick, and other controllers. It also incorporates an audio interface with 4 inputs and 4 outputs, a MIDI interface, a microphone preamplifier, instrument inputs, and a headphone output.

Ozonic has 40 MIDI-assignable controllers that you can use to program and perform with Real-Time AudioSuite (RTAS) virtual instruments and effects – bringing back tactile control. So, for example, you could assign the mod wheel to the cut-off frequency of your virtual MiniMoog or map Ozonic's slider bank to the drawbars of your virtual tone wheel organ.

Fig. 1.13 – Ozonic 37-key FireWire audio/MIDI interface and controller.

Ozonic connects to your computer via FireWire and also incorporates both MIDI and audio interfaces. So, if you already have a laptop or desktop, a pair of powered monitor speakers, and maybe a microphone or guitar, Ozonic completes your MIDI and audio recording system. M-Audio also offers a range of accessories including sustain and expression pedals, an affordable studio microphone, and Ableton Live software.

The MIDI keyboard has pitch bend and modulation wheels at the left and an octave selector button that you can press to play lower notes or higher notes than the default range. Above these there are several controls for output level, headphone level, input gain, and so forth. Directly above the keys, you will see two columns of buttons that let you select the preset, group, or zone, with an assignable joystick positioned between these. A row of 14 MIDI-assignable buttons runs above the rest of the keys. Above the buttons there is a small LCD screen, 8 assignable knobs, and 9 MIDI-assignable faders. You can use all these controls to give you hands-on control of any MIDI-controllable hardware devices, mixers, sound modules or effects, or 'virtual' instruments, effects, and other software.

The audio and MIDI interfaces are accessed via the back panel. Ozonic has just one FireWire socket, which can be a limitation if you want to use it with other FireWire peripherals. Normally, Ozonic takes its power from your computer via the FireWire bus, but a low-voltage power input is provided that you can use with any suitable mains adapter as necessary. MIDI in and out sockets are provided along with a pair of jack sockets for sustain and expression pedals. There are four 1/4″ audio output jacks, three 1/4″ audio input jacks, an XLR socket with associated phantom power switch that you can use to hook up any stage or studio microphone – and a headphone socket.

You get a cut-down version of Reason free with Ozonic to get you started. You also get a software control panel that provides basic audio mixing and metering facilities. And now that Avid/Digidesign owns M-Audio, you can buy Pro Tools M-Powered software to use as the software front-end for Ozonic. This means that the files you record using Ozonic at a gig, in the field, or on the road can be session-compatible with full Pro Tools systems at professional studios around the world.

FireWire 410 & FireWire 1814

The FireWire 410 has 2 balanced analogue inputs and 8 balanced analogue outputs plus 2 channels of S/PDIF I/O. You might choose this interface when you need to run discrete output channels to independent mixer channels and other outboard gear, or to provide separate outputs for surround monitoring. It has MIDI in and out sockets, direct hardware monitoring, and is bus-powered with an optional low-voltage power input.

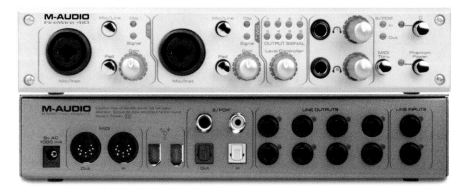

Fig. 1.14 – FireWire 410 4-In/10-Out FireWire mobile recording interface.

If you're recording a full band, the FireWire 1814 with its 8 balanced analogue inputs and 4 balanced analogue outputs makes a better choice (especially if you get the optional 8-channel Octane preamp). It also has 8 channels of digital input and output via optical 'Lightpipe' connectors, along with word clock synchronization for seamless communication with devices, such as ADATs, DA-88s, and digital mixers.

Fig. 1.15 – FireWire 1814 18-In/14-Out FireWire audio/MIDI interface w/ADAT lightpipe.

Fig. 1.16 – FireWire 1814 rear panel.

The FireWire 1814 and FireWire 410 both support up to 24-bit/96 kHz with Pro Tools M-Powered. (When used with software supporting even higher data rates, the FireWire 410 can output 2 channels at 192 kHz, and the FireWire 1814 goes so far as 2 × 4 analog at 192 kHz.) Both units have great dual mic/instrument preamps, as well as dual simultaneously active headphone amps with convenient individual front-panel controls and 1/4″ outputs.

These compact studios literally fit in a backpack (M-Audio even make the backpack) and are amazingly powerful and convenient for remote recording. Again, the files you record at a gig, in the field, or on the road are session-compatible with full Pro Tools systems at professional studios around the world.

FireWire Audiophile

Created with the laptop-based DJ or live performer in mind, the FireWire Audiophile is a compact, FireWire-compatible audio/MIDI interface in a mobile package that's perfect for live performance, home recording, and digital DJ work.

Stereo headphone output with A/B switching between assignable sources allows precuing for DJs and mobile musicians – and the assignable aux bus is perfect for creating dedicated headphone mixes and effect sends. You also get zero-latency hardware direct monitoring, ultra-low latency ASIO software direct monitoring, and MIDI I/O. The FireWire Audiophile even has AC-3 and DTS pass-through for surround sound playback.

The front panel has a 1/4″ jack headphone output with associated volume control and an A/B source switch, a pair of LEDs to indicate S/PDIF I/O activity, an assignable level control, and a power on/off button.

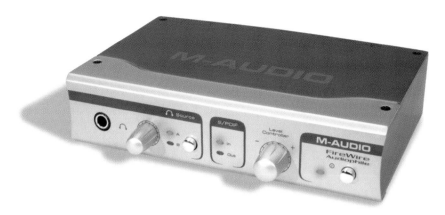

Fig. 1.17 – FireWire Audiophile 4-In/6-Out FireWire audio/MIDI interface.

The rear panel has a 12-volt low-power input, a pair of FireWire sockets, a pair of coaxial connectors providing 2 channels of S/PDIF digital input and output, a pair of MIDI in and out sockets. It also has two pairs of RCA connectors providing analogue line outputs and one pair of RCA connectors providing analogue output.

Fig. 1.18 – FireWire Audiophile rear panel.

That makes the FireWire Audiophile a four-input, six-output audio interface capable of high-quality analog and digital I/O with full 24-bit resolution at sampling rates up to 96 kHz for recording and playback.

FireWire Solo

FireWire Solo is a mobile audio interface designed for songwriters and guitarists.

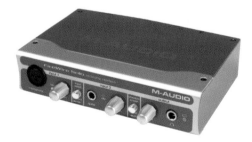

Fig. 1.19 – FireWire Solo.

The front panel has an XLR microphone input with associated gain control, a 1/4″ jack guitar input with associated gain control, a front/rear input selector, a phantom power on/off switch for the microphone preamplifier, and a head-phone output 1/4″ jack with associated volume control.

Fig. 1.20 – FireWire Solo rear panel.

The rear panel has 2 FireWire sockets, a 12-volt power input, stereo S/PDIF I/O, 2 unbalanced analogue line inputs and 2 balanced analogue line outputs via 1/4″ jack sockets.

PCI cards and interfaces

Pro Tools M-Powered works with various M-Audio Delta PCI 2.2 interface cards, including the best-selling Audiophile 2496 and the next-generation Audiophile 192. Other systems available at the time of writing include the Delta 1010, the Delta 1010 LT, the Delta 44, and the Delta 66.

Audiophile 2496 & Audiophile 192

Pro Tools M-Powered works with various M-Audio Delta PCI 2.2 interface cards, including the best-selling Audiophile 2496 and the next-generation Audiophile 192.

Both of these feature 2-channel analogue balanced 1/4″ I/O, 2-channel S/PDIF coaxial digital I/O, and up to 24-bit/96 kHz operation with Pro Tools. (The Audiophile 192 also features balanced 1/4″ I/O and provides up to 192 kHz sampling rate with applications that support it.) Both devices let you use analogue and digital I/O simultaneously, providing the flexibility of configuring 4-channel I/O, monitoring analogue while outputting digital, or running a digital effects loop.

Direct monitoring on both devices allows you to hear the input without latency while recording. The Audiophile 192 provides the extra benefit of a second pair of outputs, allowing routing to a separate headphone mix or two different speaker systems.

Fig. 1.21 – Audiophile 192 high-definition 4-In/4-Out audio card with digital I/O and MIDI.

These PCI 2.2 interfaces come with control panels that give you full digital mixing control of all inputs and outputs. Another benefit is that all M-Audio products that feature digital I/O provide pass-through of surround-sound-encoded AC-3 and DTS content, allowing you to stream surround content directly to a compatible surround receiver or similar device.

Fig. 1.22 – Audiophile 2496 4-In/4-Out audio card with MIDI and digital I/O.

Both the Audiophile 2496 and Audiophile 192 have built-in 16-channel MIDI I/O, so there's no need for a separate MIDI interface. You'll also find built-in support for ASIO, WDM, GSIF2, and Core Audio for compatibility with most other applications, such as Cubase SX, Sonar, Digital Performer, and so forth.

Delta 1010

The Delta 1010 10-In/10-Out PCI/Rack Digital Recording System with MIDI and Digital I/O consists of a PCI card that connects to a 19″ rackmountable box that contains the various input and output sockets.

Fig. 1.23 – Delta 1010 10-In/10-Out PCI/Rack digital recording system with MIDI and digital I/O.

Delta 1010 LT

The Delta 1010 LT version does not have the rackmountable audio interface. Instead, the input and output sockets attach to the PCI card via multi-way connectors and 'float' around loose at the back of your computer – which is not as robust a solution, but is even more affordable.

Fig. 1.24 – Delta 1010 LT.

Delta 44

The Delta 44 Professional 4-In/4-Out Audio Card is a PCI card that connects to a small breakout box that houses the four audio input jacks and four audio output jacks.

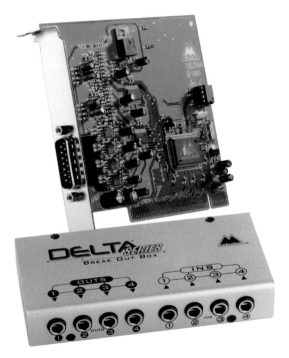

Fig. 1.25 – Delta 44 professional 4-In/4-Out audio card.

Delta 66

The Delta 66 Professional 6-In/6-Out Audio Card with Digital I/O is virtually identical to the Delta 44 but has the additional benefit of a pair of digital stereo input/output S/PDIF sockets on the back of the PCI card.

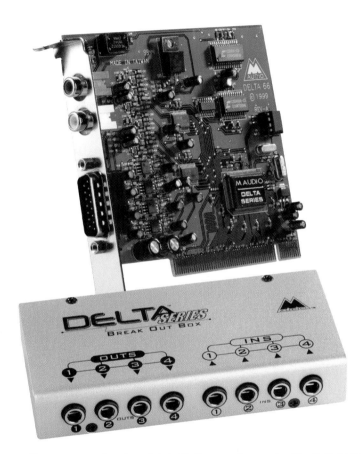

Fig. 1.26 – Delta 66 professional 6-In/6-Out audio card with digital I/O.

Latest Additions

Coinciding with the launch of Pro Tools 7, Digidesign announced that several more M-Audio devices had been qualified for use with Pro Tools M-Powered 7 software, including Black Box, Ozone, Mobile Pre USB, Fast Track USB, and Transit.

You won't be needing a crystal ball to figure out that this will be an ongoing process, so keep your eyes out for future announcements in the Digidesign and M-Audio websites and in the various music technology magazines.

2

The Digi 002 and the Command|8

The Digi 002

Resembling a typical small digital mixer, the Digi 002 actually combines the features of a high-quality audio interface, a MIDI interface, a touch-sensitive control surface, and a stand-alone digital mixer, all in one unit. Connecting to your computer through a single FireWire connection, Digi 002 pairs Pro Tools LE software with a dedicated control surface for hands-on control of the software.

Audio and MIDI data are passed back and forth between your computer and the Digi 002 via the FireWire cable along with control information generated by or returned to the Digi 002 control surface. Any moves you make using the Pro Tools software on the computer screen will be reflected on the control surface – and vice versa.

At the push of a button, the Digi 002 unit can be switched into Standalone mode to become an 8×2 digital mixer, complete with two internal and two external effects sends, EQ, dynamics, effects, reverb, and automation snapshots. So you can use Pro Tools LE with the Digi 002 to record, edit, process, mix, and master your projects in your home studio, then put the Digi 002 under your arm, take it to a gig, and use it as a digital mixer on-stage. Whoopee! It's your flexible friend!

The Digi 002 lies somewhere in that grey area between an entry-level product and a professional product that is sometimes referred to as 'prosumer'. It is definitely going to give you better audio quality than an Mbox, for example, but it is not quite in the same league as Digidesign's HD systems which offer much higher-quality converters and analogue electronics and much more choice when it comes to building systems with many inputs and outputs.

Nevertheless, the Digi 002's analogue inputs and outputs support the most important sample rates that you are likely to encounter, with A/D and D/A converters allowing sample rates of 44.1, 48, 88.2, or 96 kHz. The coaxial S/PDIF connectors support up to 24-bit, 96 kHz audio. If you use the optical connectors in ADAT mode then these can only support the original 44.1 and 48 kHz sample rates that they were designed for. But if you use these in Optical S/PDIF mode they will also support the higher sample rates of 88.2 and 96 kHz.

tip ▷ Don't forget to select the appropriate I/O format as the Clock Source in the Hardware Setup dialog when transferring material digitally into Pro Tools using ADAT or S/PDIF formats.

Rear panel

As you might expect, the Digi 002 has similar connectors on its rear panel to those on the Digi 002R.

At the lower right-hand side, there are four Microphone Inputs via balanced, three-conductor XLR connectors. If you are using condenser microphones that need to take their power from the microphone cables, you can switch on the 48-volt so-called 'phantom' power for inputs 1 and 2 or for inputs 3 and 4 using two small push-button switches located above. This 48-volt power is then supplied to the microphone via its own cable.

Also located above the microphone inputs, there are four Line/Instrument inputs that use balanced, 1/4″ TRS jacks for line-level or instrument-level inputs. These line inputs have operating levels fixed at +4 dBu, which means that they are suitable for connection to professional audio equipment. To the left of these there are four more analogue inputs that also use balanced, 1/4″ TRS jacks for line-level inputs. The operating levels of each of these inputs can be set to either +4 dBu or to −10 dBV to suit consumer audio equipment, using switches immediately to the right of the input jacks.

note ▷ A −10 dBV input pair marked 'Alternative Source' is provided using RCA connectors so that you can hook up a tape, cassette or CD player for playback.

The Digi 002 also has eight analogue audio outputs. The Main (1–2) Output pair can be connected to a professional tape or DAT machine or other equipment at +4 dBu operating level using balanced, 1/4″ TRS jacks.

Analog Outputs 3–8 also use balanced, 1/4″ TRS jacks at a fixed +4 dBu operating level – so these are intended for connection to professional audio equipment such as analogue signal processors (compressors, EQ's, reverbs, and other effects) or to headphone distribution amplifiers for musicians to use.

The Monitor Output pair mirrors the Main (1–2) Output pair and works together with its front-panel volume control to provide direct connection to a monitor amplifier or to professional powered speakers at a fixed +4 dBu operating level.

note ▷

A −10 dBV 'Alternative Main' unbalanced output pair that mirrors the Main (1–2) Outputs can be used to provide a direct connection to −10 dBV devices such as consumer-quality cassette players or to a consumer hi-fi amplifier or powered speakers if you don't have professional monitoring equipment.

At the lower left-hand side you will find a pair of optical connectors for eight channels of ADAT I/O or two channels of Optical S/PDIF I/O; a pair of RCA connectors for two channels of S/PDIF digital I/O; and two FireWire ports – one to connect to your computer and one to let you daisy-chain other FireWire devices such as digital cameras or camcorders.

To the left of these, there is a footswitch jack that you can use to punch in and out of record. To the right there are three five-pin DIN MIDI sockets – one MIDI In, and two MIDI Outs – allowing for 16 channels of MIDI input and 32 channels of MIDI output.

Finally, the power connector will accept AC supplies between 100 and 240 volts running at either 50 or 60 Hz – so you can connect the Digi 002 to the main electricity supplies in most countries around the world without problems.

The control surface

The Digi 002 control surface is divided into two main sections with the input gain and output level controls at the top and the other controls below.

The main section is further sub-divided to group the control knobs, more switches and the 'scribble strip' displays in the middle section with the moving faders below these in the lowest section and the transport controls to the right of these.

Fig. 2.1 – Digi 002 control surface.

Top section

Let's look at the top section first. At the top left there are four rotary gain controls for the four microphone preamplifiers. Above these there are four pairs of buttons. The first of each pair switches the input between microphone and line level. The second button switches in a high-pass filter. This filter removes the lower frequencies that may contain rumble or other unwanted sounds.

Fig. 2.2 – Digi 002 top section.

To the right of these there are two small button switches. The first of these routes the Alternative Source to input pair 7–8. The second routes the Alternative Source to the monitor outputs. An associated rotary control knob lets you adjust the output level to the monitors, and adjacent to this a button switch is provided to let you mute the monitors. There is also a Mono button that can be very useful when checking audio that will be broadcast (which may be in mono).

Finally, at the far right of the panel, there is a 1/4″ jack socket for headphone output with an associated volume control knob.

Middle section

Fig. 2.3 – Digi 002 middle section.

Home view

The Digi 002's default view is called the Home View. The idea here is that you should always go back to this 'home' view after going outside the 'home' view to other views.

Home view is the Console View with the Pan controls displayed on the rotary encoders and the mixer channel names shown in the Channel Scribble Strips. This makes sense as the default view because the faders control volume and the rotary encoders control pan and both of these are positioned on the control surface just where you would expect volume faders and pan controls to be on a mixer.

Console View Selectors

In the middle section, there is a row of three buttons at the far left to let you select the Console View. You can use these to choose whether to view the pans, sends or inserts in the 'Scribble Strips' located to the right of these.

The default setting is the Pan View in which, when you first launch a Pro Tools session, the faders control track volume and the rotary encoders control channel pan positions.

When you switch to Send View, the Channel Scribble Strips show the names of any currently assigned sends and the rotary encoders are assigned to control the send levels.

When you switch to Insert View, the names of any plug-ins currently assigned as inserts (whether hardware inserts or plug-ins) will appear in the corresponding Channel Scribble Strips. If you select any of these inserts in the Pro Tools Mix or Edit windows, its name will flash on and off in the Scribble Strip so that you know where it is. Select the one you are interested in by pressing the corresponding Channel Select button on the Digi 002 and the controls will be shown in the Channel Scribble Strips and assigned to the rotary encoders ready for you to adjust.

Insert/Send Position Selectors

Pro Tools has ten sends and five inserts available for each channel so the Digi 002 needs a way to switch its displays between these. At the far left of the middle section, underneath the Console View Selectors, you will find a vertical column of buttons labelled A to E. These are the Insert/Send Position Selectors and they are there to let you choose which of the available send or insert positions are displayed in Console View.

Channel Scribble Strips

The eight Channel Scribble Strips display channel information such as track name, or pan, send, or plug-in values. They can also display the fader values in Decibels (dB's).

When you move a fader or a rotary encoder, the scribble strip will temporarily display the value for that control before returning to the default display. So with the pan values showing in the scribble strips, if you move a fader, it will show the fader value while you are moving the fader and for a fraction of a second after you let go of the fader – then it will revert to showing the pan value.

Display Scribble Strips and Display Mode Switch

To the right of the Channel Scribble Strips, two more 'scribble strips' show additional information – the Display Scribble Strips.

For example, if you have any stereo tracks in your Pro Tools Session, the Channel Scribble Strips will default to displaying the pan values for the left channels. To view and be able to modify the right channel pan settings, you have to press the button marked 'L R Meter' located to the right of the encoder displays. Three associated LEDs indicate whether you have selected the left or right channels for panning, or the left or right Meter. This information is also shown in the Display Scribble Strips.

By default, the Display Scribble Strips always show the current status of the Channel Scribble Strips. So, for example, if you hit the Insert View button, it will say which insert you are viewing, such as Insert A.

If you press the Command switch at the same time as the Display Mode switch, the channels will display numerical parameter values rather than control names. For example, in Pan view, when you press Command + Display, fader volume levels in dB are shown as the default display.

You can also show the Pro Tools main Counter in the Display Scribble Strips by pressing the Display Mode switch, which is located right between the Display Scribble Strips and the Channel Scribble Strips. So you can see the Bar:Beat location or the Society of Motion Picture and Television Engineers (SMPTE) location or whatever the main Counter is set to.

More buttons and status indicators

There are various LEDs and buttons located below the Display Scribble Strips. Four LEDs indicate the sampling rate: 44.1, 48, 88.2, or 96 kHz.

Three buttons underneath these let you select Enter, Undo, and the Standalone mode.

Three more LEDs to the right of these indicate MIDI input and output activity.

To the left you will find the Record Enable button. When this is pressed you can arm any track for record by pressing its Channel Select button on the Digi 002.

Rotary encoders

Underneath the Channel Scribble Strips you will find eight rotary encoders that you can use to adjust the pan, send, meter, and plug-in channel settings. Each rotary encoder has a circle of 15 LEDs above it to indicate data values controlled by the encoder.

The style of display depends on the type of data. For example, discrete or stepped information such as pan position or frequency value is shown by a single LED, while an expanding series of LEDs shows values such as send levels, gain, or filter bandwidth. The LED rings can also be set to show track levels by pressing the Encoder/Meter mode switch to the right of the encoder area (the button marked 'L R Meter'). When you set this to Meter mode, the LED rings show increasing levels in a clockwise manner. When the last red LED lights, this indicates clipping.

Channel Select

Underneath the rotary encoders you will find a row of eight Channel Select button switches.

In Pan view, pressing a Channel Select switch selects the corresponding track in Pro Tools – which is what you would expect it to do.

In Send View, the Channel Select switch toggles the selected send on that track between pre- and post-fader operation – which is a useful function in this view.

In Insert View, pressing a Channel Select switch directly under a plug-in name does something much more radical: it puts Digi 002 into Channel View and displays the plug-in controls across all the channel strips.

Channel View

Above the Channel Scribble Strips there is a row of eight buttons. The first four of these are the Channel View Selectors that let you choose between viewing the EQ, Dynamics, Insert, or Pan/Send settings.

There are two Page buttons to let you page forward or backward when you are using plug-ins with more controls than will fit on the eight scribble strips. There is also a Master Bypass button that works with the currently selected plug-in or with all plug-ins on a channel, depending on the current view. Finally, there is an Escape button that lets you cancel certain operations, such as the Channel View selections. It also functions as a 'Cancel' button for on-screen dialogs in Pro Tools.

Channel View lets you zoom in on a single track and display all the plug-in assignments, insert names, or send assignments horizontally, across all the Channel Scribble Strips. From this view, you can recall and edit parameters for all the sends on a single track, or all the parameters of a single plug-in.

In Channel View, the LED rings above the rotary encoders indicate values for the selected control showing plug-in parameters, send levels, insert levels, or pan values depending on which Channel View switch is lit.

Don't forget that, just as in Console View, the Digi 002's faders mirror the Pro Tools software's volume faders when in Channel View.

note ▷ Dedicated buttons for EQ and Dynamics are provided because these are used so often. Any plug-ins can be accessed using the Insert button (including EQ and Dynamics) but it is more efficient to use the dedicated buttons for EQ and Dynamics if these are what you want to get to.

tip ▷ To cycle through all the plug-ins or inserts on a channel, hold the EQ, Dynamics or Insert button and repeatedly press the track's Channel Select switch.

note ▷ On Digi 002 (and Command|8), when you use Channel View to display the sends on a single Pro Tools channel, you can only view Sends A–E. To view Sends F–J, use Console View. Press the Send switch to put Digi 002 (or Command|8) into Sends View, then hold the Shift/Add switch and press the corresponding Send Position switch (A = F, B = G, C = H, D = I, E = J).

EQ   When you press the EQ switch, the Channel Select buttons will light up on any channels that have an EQ plug-in inserted so that you can identify these. Choose the one you want to work with by pressing its Channel Select button.

Controls for the first EQ plug-in on that track are then assigned to the rotary encoders and displayed in the Channel Scribble Strips and switched controls, such as Master Bypass or Phase Invert functions, can be controlled using the Channel Select switches.

If you have more than one EQ plug-in inserted on a track, you can get to the second by holding the EQ button and pressing the track's Channel Select button. The controls will then be displayed on the Channel Scribble Strips, and if the first EQ plug-in's window is open in Pro Tools, the window will switch to the second EQ plug-in.

Dynamics   When you press the Dynamics switch, the Channel Select buttons will light up on any channels that have dynamics plug-ins (such as compressors or limiters) inserted so that you can identify these.

Choose the one you want to work with by pressing its Channel Select button. Controls for the first Dynamics plug-in on that track are assigned to the rotary encoders and displayed in the Channel Scribble Strips. Switched controls, such as Phase Invert or In/Out for EQ bands, can be controlled using the Channel Select switches below the corresponding Scribble Strips. Again, if you have more than one Dynamics plug-in inserted on a track, you can get to the second by holding the Dynamics button and pressing the track's Channel Select button.

Inserts   When you press the Inserts switch, the Channel Scribble Strips will show abbreviated names for any inserts on the track and the Channel Select buttons will light up on any channels that have a plug-in (or any hardware I/O) inserted so that you can identify these.

You should check the Display Scribble Strips at this point to see which of the five insert slots you are viewing in the Channel Scribble Strips. These are labelled A to E and you can select which to view using the Insert/Send Position Selectors at the middle left of the control surface.

You can choose which insert to edit by viewing each of the five insert slots in turn until you find the insert you are looking for, then pressing the corresponding illuminated Channel Select button.

When you press the Pan/Send switch, Digi 002 identifies channels with sends assigned to them by illuminating their Channel Select switches. If no sends are present, no Channel Select switches will be lit. If you do have any sends in your Pro Tools Session, you can access the controls for any of these by pressing the corresponding illuminated Channel Select switch.

Channel Scribble Strip 1 and the first rotary encoder show track pan position for that track. (If the track is a stereo track, you can toggle between left and right pan by pressing the Encoder Mode switch immediately to the right of the encoders. This switch is marked 'L R Meter' on the Digi 002.)

Channel Scribble Strips 3–7 show the names and their rotary encoders control the levels for the first five sends on that track. (Channel Scribble Strip 2 is inactive.) In this mode, the Channel Select switches on channels 3–7 toggle pre- and post-fader metering for the corresponding send.

Multi-mono plug-ins

When working with a multi-mono plug-in, you can toggle the view between the left and right sides of the plug-in by holding down the Display Switch when you press the Channel Select switch to select a plug-in from Channel view.

The resulting display shows 'left' and 'right' in the Scribble Strips, allowing you to choose between the two sides of the multi-mono plug-in by pressing the corresponding Channel Select switch.

Lower section

The lower section contains the faders along with the transport and navigation controls.

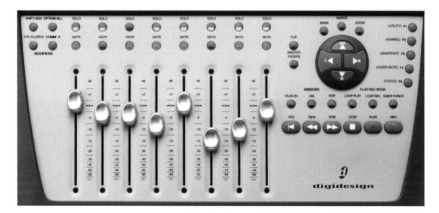

Fig. 2.4 – Digi 002 lower section.

Faders

The fader section consists of eight identical channel strips, each with a touch-sensitive motorized fader, solo and mute switches.

Transport controls

Located to the right of the fader section there is a group of transport control buttons: Play, Record, Stop, Fast Forward, Rewind, and Return to Zero.

Navigation & zoom

Above the transport controls there is a raised circular controller containing left/right and up/down navigation keys.

Above this controller there are three button switches that you can use to define the function of the left/right arrow.

Use the Bank switch to swap the Pro Tools tracks that are displayed on the Digi 002 for the next bank of eight or the previous bank of eight. If you want to move the tracks one at a time, use the Nudge switch instead.

If you press the Zoom button, you can also use the controller to control the zoom function in the Pro Tools Edit window. When this button is lit, the Left and Right arrow keys zoom the display horizontally and the up and down arrow keys zoom the display vertically in and out.

The Left and Right arrow keys can also be used to navigate between editable fields when you are editing numerical values such as Selection Start, End, and Length or Pre-and Post-Roll in the Edit or Transport windows. The Up and Down keys can be used to increment or decrement the selected value.

In Bank and Nudge modes, the Navigation keys perform the same function as the Up or Down keys on the computer keyboard. So you can use these keys to mark in and out points during playback to make selections in the Pro Tools Edit window. If you have already made a selection in the Edit window, the Up and Down keys will move the selection up and down your track list.

Window show/hide and Playback Mode switches

In between the navigation controller and the transport controls there are six more control buttons. The first, marked 'Plug-in', opens or closes the window of the currently selected plug-in The second, marked 'Mix', opens, brings forward, or closes the Mix window in Pro Tools and the third, marked 'Edit' does the same for the Edit window. The next three buttons let you switch the various Play/Record modes on or off: Loop Playback, Loop Record, and QuickPunch.

Fader Flip & Master Faders

To the left of the navigation controller there are two buttons, marked 'Flip' and 'Master Faders'.

The Fader Flip switch transfers control assignments from the rotary encoders to the corresponding channel faders, allowing you to use the touch-sensitive faders to edit and automate these control values. So, for example, if you hit the Flip button in Console View, it will move the send level controls to the channel faders and the send pan controls to the rotary encoders. Or in Channel View, if you are working with a plug-in insert, pressing the Flip button will move the plug-in's controls from the rotary encoders to the faders.

Pressing the Master Fader switch arranges any and all of the Master Fader tracks in the current session on the right-hand side of the control surface so that you can focus on making adjustments to these. Pressing this switch a second time returns the control surface to the previous view.

Function/utility keys

To the right of the navigation controller, there are five function keys labelled F1 to F5. F1 takes you into Utility Mode when the Digi 002 is in stand-alone mode so that you can set the control surface and input preferences and run diagnostic tests. Also for use in stand-alone mode, F2 lets you name channels and F3 lets you store and recall up to 24 mixer configurations.

F4 temporarily prevents the faders from moving while you are working with Pro Tools so that you can listen to your audio playback without hearing the noise of the faders moving on the Digi 002. When you want the faders to move again, just press F4 a second time. Don't worry about this affecting the sound in Pro Tools though: this feature stops the Digi 002 faders from moving but it doesn't affect the virtual faders in the Pro Tools Mix window (these will continue to move).

F5 is a short cut that lets you display the currently active plug-in's controls in the Digi 002 Channel Scribble Strips. Using this is much faster than pressing the Insert button to enter Channel View then pressing the channel select button for the plug-in you want to make active and again to display its controls in the Scribble Strips. And when you have adjusted the controls, another press on the F5 button takes you back to the view that you were in previously.

Keyboard modifier switches

To the left of the fader section there are four modifier keys corresponding to the modifier keys on your computer keyboard. These are labelled the same way as the modifier keys on Apple computers, so you have Shift, Option, Control, and Command. You can use these in the same way as those on the computer keyboard in combination with other key presses on the computer keyboard or with mouse clicks.

The Shift/Add Switch allows you to extend a track selection or add to a group of selected items. The Option/All Switch applies an action or command to all the tracks in a Pro Tools session. So, for example, this is very useful if you want to insert an EQ plug-in onto every track: just hold the Option key while you are inserting the plug-in onto the first track and it will insert it onto every track of the same type. The Control/Clutch Switch temporarily stops a control from acting as part of a group of controls. If you need finer adjustment of any Pro Tools controls or automation breakpoints, simply hold down the Command Switch while you adjust these and small increments will become possible.

Digi 002 modes

The Digi 002 goes into Standby mode when you power it up and it waits in this mode until you either launch Pro Tools LE or switch to Standalone mode. When the Digi 002 is connected to a computer with Pro Tools LE software running, it goes into Pro Tools mode.

The Digi 002 can also be used as a stand-alone eight-channel digital audio mixer – with or without being connected to the computer. You can switch to Standalone mode either from Standby mode or from Pro Tools mode and your computer is no longer needed for Digi 002 to operate.

Standalone Mode

To put the Digi 002 into Standalone mode just press the Standalone switch on the right-hand side of the Digi 002 top panel and confirm that you want to enter Standalone mode by pressing the Channel Select switch that is flashing under the word 'Yes'. To exit Standalone mode, simply press the Standalone switch a second time. Digi 002 goes into Standby mode, or re-enters Pro Tools mode automatically if Pro Tools software is still running.

In Standalone mode, the Digi 002 has dedicated, in-line three-band EQ on input channels 1–8 and dedicated, in-line compressors on input channels 1–4. There are four sends available on each channel for adding internal Delay or Reverb effects, or for hooking up external effects processors. Sends 1–2 are dedicated to the internal Delay and Reverb effects. Sends 3–4 route the input signals out via outputs 7–8 on the back panel so you can send to external effects.

The ten 'Scribble Strips' display the pan, volume and effects controls, send levels, and track names. The LED Encoder rings on each channel strip show the pan positions in Pan View and Master Fader view but show send levels or other information in other views. If you press the button marked 'L R Meter' this puts the Digi 002 into Meter mode so that the LED rings act as post-fader meters unless you are viewing the compressor controls, in which case they act as input, output, and gain reduction meters for the displayed compressor.

Hooking up external sources to the Digi 002 in Standalone mode

Inputs 1–4 accept microphone-, line-, or instrument-level signals and you can adjust the input gains for these using the controls located at the left of the Digi 002's top panel. Inputs 5–8 only accept line-level signals and are switchable between $-10\,dBV$ and $+4\,dBu$ using the operating level switches on the Digi 002 back panel.

The Alt Src Inputs can be used to hook up alternative audio sources such as CD players or tape decks. These can either be routed directly to the Monitor and Headphone outputs by pressing the Alt Src to Mon switch or they can be routed to Input channels 7–8 by pressing the Alt Src to 7–8 switch (in which case inputs 7–8 on the back panel of the Digi 002 are disabled).

If you have a DAT or CD player with S/PDIF digital outputs you can route these via the Digi 002's S/PDIF digital connector to inputs 5–6, allowing you to bring a stereo digital signal into the Digi 002. When playing audio into the Digi 002 digitally, you will need to change the sync mode settings to S/PDIF.

tip ▶ Some digital audio equipment (and even some personal computers such as the Mac G5) features Optical S/PDIF inputs and outputs that you can connect directly to the Digi 002. This can be useful when you want to play iTunes back through your main monitors, or to set up iChat on your Mac for video conferencing over the Internet using your main monitors instead of the Mac's internal speakers. Don't forget to set the sync mode to Word Clock, set the S/PDIF inputs to Optical, and set up the routing for the S/PDIF inputs to Digi 002 inputs 5–6 first. You can access these settings in Standalone mode by pressing the F1 Utility button, choosing Preferences by pressing the corresponding Channel Select button, and pressing the appropriate Channel Select buttons to access the desired settings.

In Standalone mode, all eight Input channels, the Delay return, and the Reverb return are summed to outputs 1–2, which are routed to the Main Outputs, to the Alt Main Outputs, and to the S/PDIF outputs on the back panel of the Digi 002. Outputs 1–2 are also mirrored on the Monitor Outputs and Headphone Output.

note ▷ If Optical is chosen in the S/PDIF preferences, only Main Outputs 1–2 are mirrored in the Optical Output port. If RCA is chosen in the S/PDIF preferences, all eight Input channels are passed directly to the eight ADAT Optical Outputs, pre-fader, pre-effects, except for the high-pass filter on channels 1–4. This allows you to route input signals directly to an ADAT device without repatching cables.

Digidesign Command | 8 Control Surface

It can be very useful to have a hardware control surface to use with Pro Tools – especially if you do not have an audio mixer. The Command | 8 is similar to the Digi 002 – but without the audio interface features. It is primarily a control surface – although it does have basic monitoring and routing facilities.

note ▷ You can use the Command | 8 to provide extra faders with a Digi 002 and it can also be used for remote control of play and record functions with Digidesign's ProControl and Control | 24 professional control surfaces.

Fig. 2.5 – Digidesign Command | 8 control surface with 002R interface and laptop.

Designed specifically for use with Pro Tools, the Command | 8 Control Surface has a big, bright 110-character backlit LCD display that is at least twice as nice as the Digi 002's display and connects to your computer using a standard USB cable. It has eight sets of touch-sensitive moving faders and rotary encoders that you can easily shift in banks to control all of the tracks in your Pro Tools session.

The Command | 8 gives you full control of all the channel strip functions in Pro Tools, allowing you to view and edit plug-in parameters and automate all the sends, pans, track volume, and mutes. Transport controls are also included, along with a footswitch jack that can be used to punch into record.

Fig. 2.6 – Command | 8 top showing details of controls.

The Command | 8 also has one MIDI In and two MIDI Out sockets and can work as a MIDI controller with any third-party MIDI software and devices that will allow you to map MIDI control change messages for level, panning, solo, mute, MIDI Machine Control, and other parameters to Command | 8 controls.

Fig. 2.7 – Command | 8 back.

You can connect the analogue audio stereo mix outputs from an Mbox, Digi 002 Rack or any of the M-Powered interfaces to the Command | 8's Main Monitor inputs and route this audio via the Command | 8's Speaker Outputs to a pair of powered monitors for control room monitoring.

A pair of External Source inputs lets you hook up another stereo source such as a CD player. The inputs and outputs all use balanced 1/4" TRS jacks and are individually switchable between −10 dBV and +4 dBu operating levels. A switch is provided to let you choose between Main and External input sources, along with control room Level, Mono and Mute switches. A Headphone output is also provided with its own separate level control.

3

Getting to Know the Pro Tools LE Software

So what are the main features in Pro Tools LE and how should you get familiar with these? You will need to know about the various Edit Modes, how to zoom the display, which Tool to use for which task, how the rulers work, how scrolling works, and how to find your way around the software and the waveform display in general. Working 'hands-on' with the software as much as possible is the best way to learn how to use any software, of course. But you can use this chapter to get an overview of the software environment, familiarize yourself with the main windows, and take a peek inside all the menus.

Pro Tools LE Overview

Pro Tools LE software lets you record and play back up to 32 mono digital audio tracks, depending on your computer's capabilities, and has comprehensive editing and mix automation features. Sessions can include up to 256 MIDI tracks, 128 Auxiliary Input tracks, 64 Master Fader tracks and 128 audio tracks – although only 32 audio tracks can be played back at the same time. But other tracks in your Session will play when any of the first 32 tracks have no audio playing. The higher ordered tracks take precedence over the lower ordered tracks, so if the first 32 tracks all have audio playing throughout the Session, no other tracks will be able to play. But if some of these tracks have gaps containing no audio regions, then a corresponding number of lower-ordered tracks will be able to play audio in those places. You can have up to five inserts and ten sends per track and you can use up to 32 internal mix busses. Sessions can use 16-bit or 24-bit audio resolution, at sample rates that depend on the capabilities of your hardware.

The Edit and Mix windows are the main Pro Tools work areas. Depending on which phase of your project you are in or what type of project you are working on, you may prefer to work with just the Mix or just the Edit window.

tip ▷ You can show or hide the Mix and Edit windows using the Windows menu or by pressing Command-Equals (=) on Macintosh, or Control-Equals (=) in Windows to switch between the two windows.

The Transport controls are available in a separate window or can be revealed in the Edit window by selecting them in the View menu.

Fig. 3.1 – Revealing the Transport controls in the Edit window.

Track Priority and Voice Assignment

Pro Tools LE and M-Powered restrict the number of playback 'voices' to 32, although the Pro Tools software allows for additional audio tracks beyond that fixed number of voices – up to a maximum of 128.

When the number of tracks exceeds the number of available voices, tracks with lower priority may not be heard. Tracks with higher positions (leftmost in the Mix window or topmost in the Edit window) have priority over tracks in lower positions in a session. To change a track's priority all you need to do is to drag the track by its Track Name button until it is lower or higher in the Edit window or to left or right in the Mix window. Alternatively, you can drag the track name in the Show/Hide Tracks List to a higher position in the list. Tracks at the top of this list have higher priority than those below.

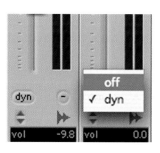

Fig. 3.2 – Voice Selector popup has two settings: 'dyn' or 'off'. Here, the left track is set to 'dyn' and the right track is about to be set to 'off'.

You can also adjust the relative priority of tracks by freeing up the voices of some of the tracks, making these available to other tracks in the session. One way to free up the voice of a track is to set the track to have 'no output' and remove any send assignments. A faster way is to click and hold the Voice Selector popup of the track and set this to 'off'.

note ▶ Pro Tools LE automatically takes care of voice management in the background, dynamically allocating voices as necessary. Each track's voice assignment can be turned off or can be dynamically allocated. When you turn a track's voice assignment off, this voice is automatically re-allocated to another track if needed; in other words it is 'dynamically allocated'.

A third way to free up a track's voice is to totally deactivate the track. There are two ways to do this. If you click and hold the Track Type icon, a popup selector appears that lets you make the track inactive when you select and highlight this. The Track Type icon on an audio track, for example, looks like a double arrowhead pointing to the right.

tip ▷ An even faster way to deactivate the track is by holding the Command and Control keys (Start and Control keys for Windows) while clicking its Track Type icon in the Mix window.

Fig. 3.3 – Click and hold the track icon to reveal a popup selector that lets you make the track inactive when you select and highlight this – or simply Command-Control-click on the icon.

The Mix Window

The Mix window lets you balance all your levels, pan the sounds to create a stereo effect, and apply signal processing as needed. There are five insert 'slots' available for each track. Each slot can be used either to hook up external processors (an outboard EQ, compressor, or whatever) or to insert a software plug-in. A range of software plug-ins is provided with Pro Tools LE and you can buy additional plug-ins from Digidesign or from third party suppliers. The Mix window also features comprehensive automation facilities, including both 'snapshot' and real-time automation with four real-time automation modes: Read, Touch, Latch, and Write.

Plug-ins

There are two formats of plug-ins that can be used with LE and M-Powered systems. Real-Time AudioSuite (RTAS) plug-ins can be inserted into the Pro Tools mixer to process audio in real time using the host computer's CPU.

AudioSuite plug-ins can be used for non-real time processing of audio regions to create new files.

The more RTAS plug-ins you use in a session, the greater the impact this will have on aspects of your system's performance such as maximum track count, number of available voices, the density of edits possible, and latency in automation and recording.

To allow creative processing, the list of RTAS plug-ins includes D-Verb, Mod Delay II, EQ III, EQ II, Dynamics II, and Pitch. Other RTAS plug-ins are provided for more technical tasks, including Signal Generator, Time Adjuster, Trim, Click, Dither, and POW-r Dither.

AudioSuite plug-ins are used to process and modify audio files on disk, rather than non-destructively in real time. Depending on how you configure a non-RTAS plug-in, it will either create an entirely new audio file, or alter the original source audio file.

For creative processing, the list of AudioSuite plug-ins includes EQ III, EQ II, Dynamics II, D-Verb, Chorus, Flanger, Multi-Tap Delay, Ping-Pong Delay, Time Compression/Expansion, Pitch Shift, and Reverse. More technical processes include Invert, Duplicate, Delay, Normalize, Gain, Signal Generator, and DC Offset Removal.

Tracks

In the Mix window, tracks appear as mixer channel strips with controls for signal routing, input and output assignments, volume, panning, record enable, automation mode, and solo/mute. From the View menu you can choose whether or not to display the Sends, Inserts, and Instrument controls in the Mix (or Edit) window.

Inserts

Pro Tools lets you use up to five Inserts on each audio track, Auxiliary Input, Instrument Track, or Master Fader. Each insert can be either a software plug-in or an external hardware device.

Sends

You can also use up to ten Sends on each track. Sends let you route signals across internal busses or to audio interface outputs so that one plug-in or one external signal processor can be used to process several tracks at once. There are two sets of sends (labeled A–E and F–J) that can optionally be displayed in the Mix (and Edit) window. You might use the first set to send to effects (such as reverb) that you wish to apply to several tracks, and use the second set to send cue mixes to musicians – routing these from your Pro Tools hardware interface to suitable headphone amplifiers.

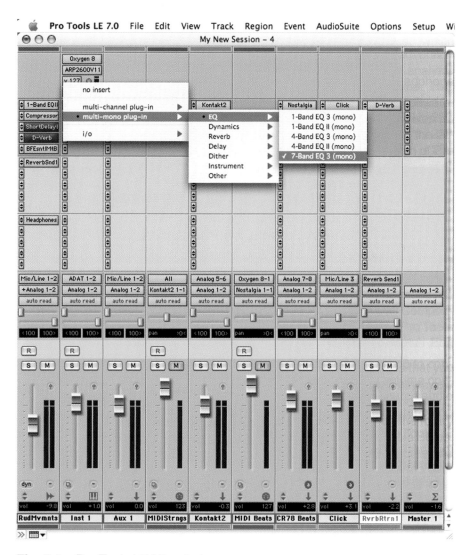

Fig. 3.4 – Pro Tools LE Mix window.

I/O Selectors

Track Input and Output Selector popups are located just above the Pan controls on each channel strip. The Track Input Selector popups let you choose the input source for Audio and Instrument tracks and Auxiliary Inputs. Track input can come from your hardware interface or from an internal Pro Tools bus – or from a plug-in.

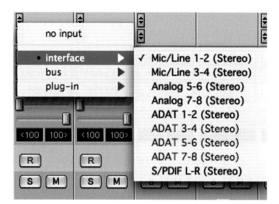

Fig. 3.5 – Track Input Selector popup.

The Track Output Selector popups let you route the audio from each track to your choice of available outputs or bus paths.

Fig. 3.6 – Multiple assignments for a Send's Audio Output Path.

note ▷ Pro Tools Audio tracks, Instrument Tracks and Auxiliary Inputs can have multiple track output and send assignments chosen from the actual paths and resources available on your system (although Master Faders can only be assigned to a single path).

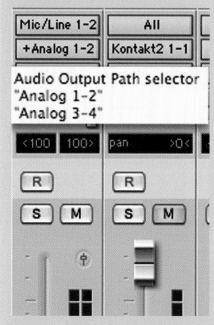

Assigning to multiple paths is an efficient way to route an identical mix to other separate outputs, for simultaneous monitor feeds, head-phone mixes, or other situations where a parallel mix is needed.

To assign an extra output, hold the Control key (Start key in Windows), open the Output Selector and select your additional output.

A '+' sign is added to the Output Selector legend to remind you that this track has more than one output assigned, and you can add as many additional outputs as there are available on your system.

If you also hold the Option (Alt) key at the same time as the Control (Start) key, the additional output will be added to all tracks (apart from Master Faders and MIDI tracks, of course).

Fig. 3.7 – Multiple assignments for a track's Audio Output Path.

tip ▷ You can use the same procedure to add additional output assignments to the track Sends.

The Edit Window

The Edit window provides a timeline display of audio, as well as MIDI data and mixer automation for recording, editing, and arranging tracks. As in the Mix window, each track has controls for record enable, solo, mute, and automation mode.

Using the View menu options, you can also reveal the input and output rout-ing assignments, the Inserts, the Sends, the Instrument controls, the MIDI Real-Time properties, the Comments, or any combination of these – which makes it possible to work with just the Edit window for most of the time.

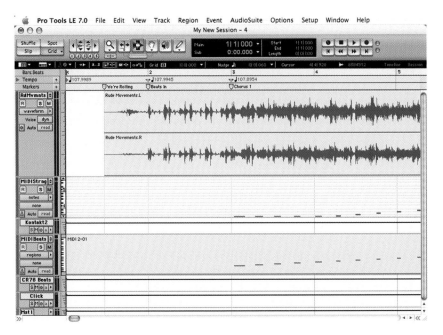

Fig. 3.8 – Pro Tools LE Edit window.

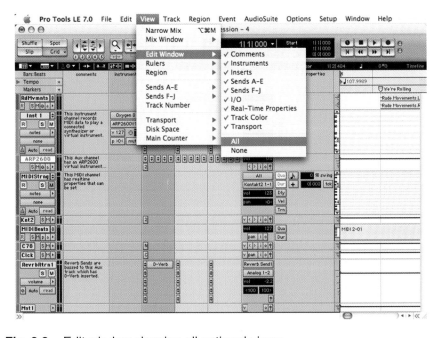

Fig. 3.9 – Edit window showing all optional views.

The Edit window lets you display the audio and MIDI data in a variety of ways to suit your purpose and you can edit the audio right down to sample level in this one window.

The Edit Modes

Understanding the area at the top of the Edit window is fundamental to knowing how to operate Pro Tools.

At the left there are four buttons to let you select the Edit Mode:

Slip Mode is the basic mode to use by default. In this mode, you can freely move regions forwards and backwards in time in the Edit window.

In Grid Mode, movements are constrained by whichever Grid settings you have made – Bars:Beats, Mins:Secs, or whatever.

In Shuffle Mode, when you move a region, it will automatically snap to the region before it – perfectly butting up to this.

And in Spot Mode, a Spot Dialog appears whenever you click on a region. This lets you specify exactly where the region should be placed on the timeline – ideal for 'spotting' effects to picture.

Fig. 3.10 – Screenshot shows the Edit Mode and Zoom buttons, the various Tools, the Location Indicators and the Event Edit Area to the right of this, with popup selectors for the Views and Rulers, Grid and Nudge values underneath.

Zoom Buttons

To the right of the Edit Mode buttons, there are various arrow buttons that let you zoom the display vertically or horizontally. The horizontal zoom arrows work for both audio and MIDI regions. To zoom audio regions vertically, use the first pair of up and down arrow buttons. To zoom MIDI regions vertically, use the second pair.

Underneath these zoom arrows there are five small Zoom Preset buttons that store preset zoom levels. You can use these as handy shortcuts to particular zoom levels: just set the zoom level you want, then Command-click (Mac) or Control-click (Windows) on any of the five buttons to store the current zoom level.

Choose Your Tool

To the right of the Zoom buttons you will find six Tool buttons. The first is the Zoomer Tool. You can use this to zoom the display either vertically or horizontally.

The next button is the Trimmer tool. Use this to lengthen or shorten regions. You can also choose to apply Time Compression/Expansion directly in the Edit window, or to apply the Scrub feature prior to trimming regions. Select these modes using the popup that appears when you click on this tool.

To the right of this is the Selector Tool that lets you use the cursor to select areas within the Edit window.

To the right again is the Grabber Tool – the one with the 'hand' icon. You can use this to move regions around in the Edit window. You can also use the Grabber to automatically separate an edit selection and move it to another location or another track using its Separation Mode.

note ▶ The Grabber Tool also offers a third mode – the Object mode – that you can enter using the popup that appears when you click on the Grabber Tool. This 'Object Grabber' lets you select non-contiguous regions on one or more tracks. Just take a look at Fig. 3.11 to see an example of a 'non-contiguous' selection of three regions, all on different tracks. By the way, 'non-contiguous regions' in this context basically means regions that are not next to each other.

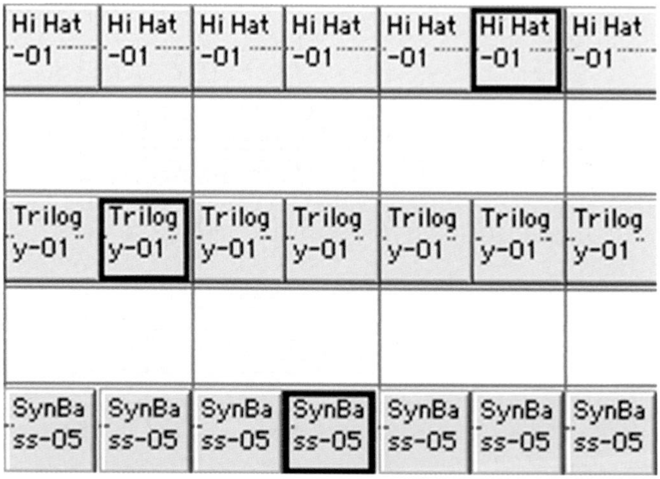

Hi Hat -01	Hi Hat -01	Hi Hat -01	Hi Hat -01	Hi Hat -01	**Hi Hat -01**	Hi Hat -01
Trilog y-01	**Trilog y-01**	Trilog y-01	Trilog y-01	Trilog y-01	Trilog y-01	Trilog y-01
SynBa ss-05	SynBa ss-05	SynBa ss-05	**SynBa ss-05**	SynBa ss-05	SynBa ss-05	SynBa ss-05

Fig. 3.11 – A Selection of non-contiguous regions in the Edit window.

tip ▶ These three tools (Trimmer, Selector, Grabber) can be combined using the Smart Tool button below these to link them together. Depending on where you point your mouse in the Edit window, one or other of these tools will become active – saving you having to click on these tools individually when you want to change to a different tool.

To the right of the Grabber Tool is the Scrub Tool that lets you 'scrub' back and forth over an edit point while you are trying to hear the exact position of a particular sound – rather like moving a tape back and forth across the playback head in conventional tape editing.

The sixth tool is the Pencil that you can use to redraw a waveform to repair a pop or click. Alternate Pencil modes are available that constrain drawing to lines, triangle, square, or random shapes.

Repairing Waveforms using the Pencil Tool

One of the most common repairs that you may want to make is to remove a pop or click from an audio file. Pops and clicks typically look like a sudden sharp spike in the waveform. You can use the Pencil tool to redraw the waveform data to remove any such unwanted spikes.

note ▷ As this form of edit permanently alters the audio file, you should work on a copy of the audio file until you become very skilled with this type of edit. A convenient way to do this is to use the AudioSuite Duplicate plug-in.

Step 1. Locate the click or pop by listening carefully and by looking along the waveform.

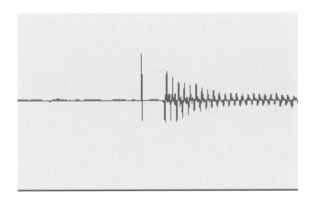

Fig. 3.12 – The sharp spike just before the wanted audio waveform is an unwanted click.

Step 2. The Pencil tool only becomes active when the Edit window is zoomed into sample level, so you need to zoom in when you have located the click or pop. Don't forget that you can always recall zoom levels using the Zoom Preset buttons or using Memory Locations.

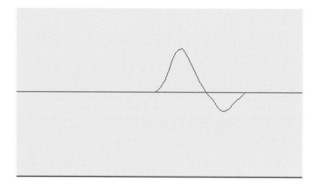

Fig. 3.13 – Zoom in until you can see the click at sample level.

Step 3. Choose the Pencil tool and use this to redraw the waveform to remove the click. It can take a little getting used to, so if you mess up at first, you can use the Undo command to undo the edit and try again.

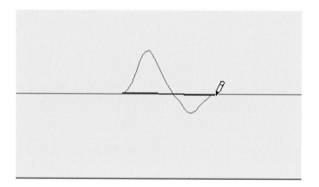

Fig. 3.14 – Redrawing the waveform.

Step 4. Audition the audio region to check that the click has been successfully removed. You may still hear a small disturbance to the sound when you solo the audio, but this should not be audible within a mix.

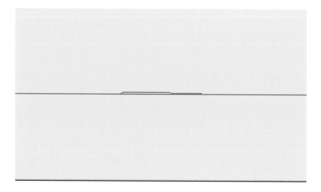

Fig. 3.15 – Waveform with click removed.

Location Indicators

Located centrally at the top of the Edit window you will find the Main and Sub Location Indicators with the Event Edit area to the right of these. The Location Indicators show you where you are in your session in terms of Bars:Beats, Mins:Secs or whatever you have chosen to display here. The Event Edit Area lets you define selections by typing the Start and End points, for example – and also serves to display these.

note ▷ If you select a MIDI note using the Grabber or Pencil Tools, an additional area appears to the right of the Event Edit area to display the selected MIDI note with its associated MIDI On and Off velocities.

Fig. 3.16 – Main and Sub Location Indicators and Event Edit Area with a MIDI note displayed at the right of the display.

Edit window controls and displays

Running underneath the Mode and Zoom buttons, Tool buttons and Location Indicators, there is a further selection of controls and displays. Here you will find popup selectors for the Views and Rulers at the left.

Immediately to the right of these you will find the Linearity Display Mode popup. This lets you choose between Linear Tick (Bars:Beats) Display and a Linear Sample (absolute) Display.

To the right again, five more buttons provide fast access to commonly used features:

The Tab to Transients button lets you automatically locate the cursor to the next transient while editing waveforms.

The Commands Keyboard Focus button enables keyboard command shortcuts.

The Link Timeline and Edit Selection button lets you link or unlink Edit and Timeline selections.

The Link Track and Edit Selection button also does what it says: with this highlighted, when you select a region in the Edit window, the track becomes selected as well. If you then select another track, Pro Tools selects the corresponding region (to the region selected in the first track) in that track – because the Track and Edit selection features are linked.

The button to the right of this lets you enable or disable Mirrored MIDI Editing. Mirrored MIDI Editing is useful when you edit a region containing MIDI notes and you want these edits to apply to every MIDI region of the same name.

To the right of these buttons there are displays for the Grid and Nudge values followed by a display area that tracks the position of the cursor. This area also displays cursor values such as the MIDI note when the cursor is moved vertically within a MIDI track or the Volume level in dB when the cursor is moved vertically within a Master track.

Keyboard Focus

Depending on which Keyboard Focus is enabled, you can use the keys on your computer's keyboard to select regions in the Region List, enable or disable Groups, or perform an edit or play command.

note ▷ There are three types of Keyboard Focus, but only one of these can be active at a time, so when you engage one Keyboard Focus it will disable the one previously engaged.

tip ▷ You can choose the Keyboard Focus by holding Command-Option (Mac) or Control-Alt (Windows) while you press 1 for the Commands, 2 for the Region List, or 3 for the Group List.

Fig. 3.17 – Commands Keyboard Focus.

Commands Keyboard Focus When the Commands Keyboard Focus is selected, a wide range of single key shortcuts for editing and playback become active on your computer's keyboard. To see these commands listed, choose 'Keyboard Shortcuts' from the Pro Tools Help menu.

tip ▷ When the Commands Keyboard Focus is disabled, you can still use the keyboard shortcuts by pressing the Control key (Mac) or Start key (Windows) at the same time as the shortcut key.

Region List Keyboard Focus When the Keyboard Focus is selected by clicking the a–z button in the Regions List, audio regions, MIDI regions, and Region Groups can be located and selected in the Region List by typing the first few letters of the region's name.

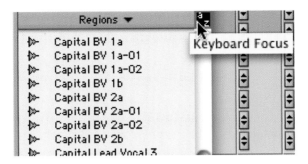

Fig. 3.18 – Enabling the Region List Keyboard Focus.

Group List Keyboard Focus
When the Keyboard Focus is selected by clicking the a–z button in the Edit Groups list, you can enable or disable the Mix and Edit Groups by typing the Group ID letter (a, b, c, etc.) on your computer keyboard when using either the Mix or Edit window.

Fig. 3.19 – Enabling the Group List Keyboard Focus.

The Rulers

At the top of the tracks display in the Edit window Pro Tools LE can show 'ruler' tracks running above the audio and MIDI tracks.

Timebase Rulers can be set to Bars:Beats, Mins:Secs, or Samples. These determine the format of the Main counter and provide the basis for the Edit window Grid.

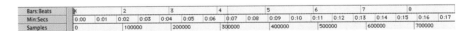

Fig. 3.20 – Timebase Rulers.

If you have DV Toolkit installed, additional timebase rulers become available for Timecode and for Feet + Frames. These can be used in Spot mode to spot audio to picture.

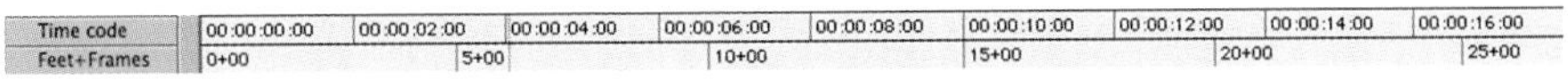

Fig. 3.21 – Time code and Feet + Frames Rulers.

Ruler tracks are also available to display Meter events, Marker events and Tempo events, and the Tempo ruler can be 'opened' to reveal the Tempo Editor.

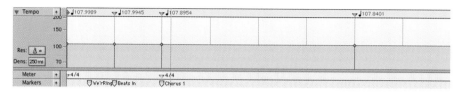

Fig. 3.22 – 'Conductor' Track Rulers for Meter, Markers and Tempo, with the Tempo Editor revealed.

The View menu lets you choose which rulers to have visible in your project, allowing you to have all 'on' or 'off' or any combination that you like.

Scrolling Options

There are three scrolling options in Pro Tools LE. The first of these, 'None', is self-explanatory. 'After Playback' leaves the window where it is when you start playback and scrolls the view to the new position after you stop playback. 'Page' scrolling moves the view a page at a time as is necessary to keep the playback cursor in view at all times.

Playback Cursor Locator

The Playback Cursor Locator lets you locate the playback cursor when it is off-screen. For example, If scrolling is not active (set to 'None'), when you stop playing back, the Playback Cursor will be positioned somewhere to the right, off the screen – if it has played past the location currently visible in the Edit window. Also, if you manually scroll the screen way off to the right, perhaps to check something visually, then the Playback Cursor will be positioned somewhere to the left off the screen.

To allow you to quickly navigate to wherever the Playback Cursor is positioned on-screen, you can use the Playback Cursor Locator button.

note ▶ The Playback Cursor Locator button only appears under certain conditions:

It will appear at the *right* edge of the Main Timebase Ruler after the playback cursor moves to any position *after* the location visible in the Edit window.

It will appear at the *left* edge of the Main Timebase Ruler if the playback cursor is located *before* the location visible in the Edit window.

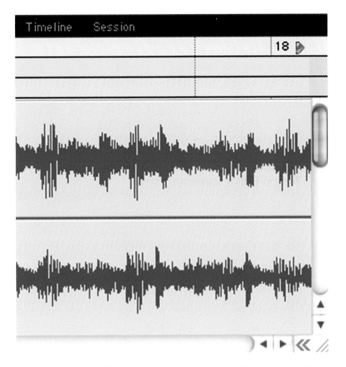

Fig. 3.23 – The Playback Cursor Locator, the small blue arrowhead, can be seen at the top right in the Main Timebase Ruler (which is displaying bar location 18 in this example).

A click on the Playback Cursor Locator immediately moves the Edit window's waveform display to the Playback Cursor's current on-screen location – saving you lots of time compared with any other way of finding this location.

note ▶ The Playback Cursor Locator is red when a track is record enabled and blue when no tracks are record enabled.

The Transport Window

In standard view, the Transport window just has a row of buttons providing controls for Online, Return To Zero, Rewind, Stop, Play, Fast Forward, Go to End, and Record.

The View menu lets you show Counters and MIDI Controls as well. The top row of MIDI controls includes buttons for Wait for Note, Metronome Click, Countoff and MIDI Merge. The Main Counter shows your choice of Bars:Beats, Mins:Secs, or Samples. (With DV Toolkit installed it can also display Timecode or Feet + Frames.)

You can also switch the Transport window to its Expanded display. This adds a Sub-counter underneath the main counter, along with Conductor, Meter, and Tempo controls in the MIDI section. A small slider lets you manually adjust the tempo or, alternatively, you can enable the Tempo track by clicking on the 'conductor' button.

Underneath the Transport controls, the Expanded window adds Pre- and Post-Roll settings, start, end, and length indicators for Timeline selection, and a Transport Master popup selector.

Fig. 3.24 – Pro Tools LE Transport Window Expanded and showing the Counters and MIDI Controls.

tip ▷ To start and stop playback, simply press the Spacebar on your computer keyboard.

Pro Tools LE Menus

One of the best ways to get familiar with software is to find out what is available in the various menus. Many of the menu items also have equivalent keyboard commands that you can use instead of selecting from the menus. Using the keyboard commands is faster, but until you have memorized these, there is no reason not to select from the menus.

tip ▷ Each time you select a menu item, notice whether there is a keyboard command that will let you do this more quickly. Over time, this will help you to memorize these keyboard commands.

Pro Tools Menu

On the Mac, there is an extra menu at the left of the menu bar. This Pro Tools LE 7.0 menu contains the Quit, Show All and Hide commands and lets you access the standard OSX Services and the Pro Tools Preferences. There is no equivalent menu to this for Windows.

Fig. 3.25 – Pro Tools Menu.

File Menu

The File menu contains the usual 'New', 'Open', 'Close', 'Save' and 'Save As. . .' file commands along with a 'Revert To Saved' command that lets you close your current session and open the last saved version. A recently added command lets you send your file via Digidesign's DigiDelivery Internet file transfer service. You can also 'Bounce' your audio to Disk or to QuickTime Movie. Lots of file Import options are provided and you can export MIDI files.

Pro Tools 7.0 lets you import a wide range of file types including MP3, REX, ACID and AAC audio, including audio with AAC, MP4, and M4a file extensions.

note ▷ Pro Tools cannot import protected AAC or MP4 files with the M4p file extension. These files are protected under the rules of digital rights management, and cannot be imported.

Fig. 3.26 – Pro Tools LE File Menu showing Import options sub-menu.

note ▷ If you have the DigiTranslator or DV Toolkit options installed, you can also export your sessions or selected tracks as OMF/AAF files.

The Import Session Data dialog lets you bring in several useful types of data from any other Pro Tools Session files. You may have a percussion track in one song that you can use again in another song, or perhaps you have a Tempo or Meter map that you want to use again, or a set of Markers/Memory locations.

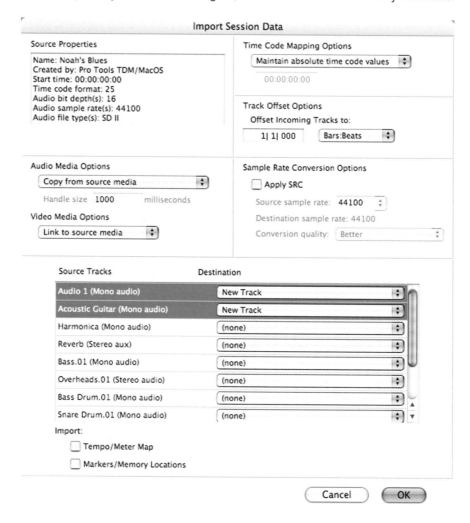

Fig. 3.27 – Import Session Data dialog.

Edit Menu

The Edit menu features a range of Cut and Paste editing commands. Some of the most useful commands, such as Repeat and Shift, Strip Silence or Insert Silence, Trim and Separate region are located here.

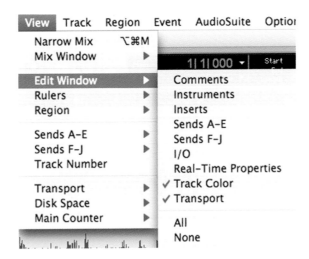

Edit	View	Track	Region	Eve
Can't Undo				⌘Z
Redo Delete MIDI Note				⇧⌘Z
Cut				⌘X
Copy				⌘C
Paste				⌘V
Clear				⌘B
Cut Special				▶
Copy Special				▶
Paste Special				▶
Clear Special				▶
Select All				⌘A
Play Edit Selection				⌥[
Duplicate				⌘D
Repeat...				⌥R
Shift...				⌥H
Insert Silence				⇧⌘E
Trim Region				▶
Separate Region				▶
Heal Separation				⌘H
Strip Silence				⌘U
Consolidate				⌥⇧3
Thin Automation				⌥⌘T
Fades				▶

Fig. 3.28 – Pro Tools LE Edit Menu.

View Menu

The View menu lets you customize the displays to suit the way you prefer to work. So, for example, you can make the mixing channels narrow so that more fit across your screen, and you can choose whether to display the Comments, I/O, Inserts and Send views independently in the Mix and Edit windows.

You can choose which Rulers to display in the Edit window and whether the Main Counter will show Bars and Beats, Minutes and Seconds, or Samples (and Time Code or Feet and Frames if DV Toolkit is installed).

Region-related commands let you choose what to display in the regions and for each set of sends you can choose whether to display all five Send assignments or just the parameters for one individual send (A–E or F–J) from the set.

View	Track	Region	Event	AudioSuite	Optior
Narrow Mix	⌥⌘M				
Mix Window	▶				
Edit Window	▶	1|1|000 ▾ Start			
Rulers	▶	Comments			
Region	▶	Instruments			
		Inserts			
Sends A–E	▶	Sends A–E			
Sends F–J	▶	Sends F–J			
Track Number		I/O			
		Real-Time Properties			
Transport	▶	✓ Track Color			
Disk Space	▶	✓ Transport			
Main Counter	▶				
		All			
		None			

Fig. 3.29 – Pro Tools LE View Menu.

Track Region Event AudioSuite

New... ⇧⌘N
Group... ⌘G
Duplicate... ⌥⇧D
Split into Mono
Make Inactive
Delete

Write MIDI Real-Time Properties

Input Only Monitoring ⌥K
Scroll to Track... ⌥⌘F
Clear All Clip Indicators ⌥C

Fig. 3.30 – Pro Tools LE Track Menu.

Track Menu

The Track menu gathers all track-related commands into one place. This is where you can create new tracks, group, dupli-cate, split, or delete tracks. Other items let you Clear All Clip Indicators, toggle Input Only Monitoring on and off, Scroll to a particular Track, and Write MIDI Real-Time Properties.

One of the most useful of these is 'Duplicate. . .' which opens the Duplicate Tracks dialog box. This lets you choose how many duplicate copies of a track to make and select which properties of the original track will be copied to the duplicates.

Properties include the active Edit playlist currently visible in the original track; the Alternate Playlists (any other hidden Edit playlists from the original track); the automation from the original track; all the Insert and plug-in assignments; all the Sends and send assignments; and the Group Assignments (all the Mix and Edit Group assignments).

Fig. 3.31 – Duplicate Tracks dialog.

Region Menu

The Region menu contains commands that you will frequently apply to selected regions, such as Mute, Lock, Capture, and Quantize to Grid.

If you are working on dance music you will frequently use the 'Loop. . .' command to open the Region Looping dialog that lets you loop MIDI or audio regions.

For example, you might have a 4, 8, or 16-bar region that you want to repeat a certain number of times, or even throughout your song.

Region	Event	AudioSuite
Mute/Unmute		⌘M
Lock/Unlock		⌘L
Send to Back		⌥⇧B
Bring to Front		⌥⇧F
Group		⌥⌘G
Ungroup		⌥⌘U
Ungroup All		
Regroup		⌥⌘R
Loop...		⌥⌘L
Unloop...		
Capture...		⌘R
Remove Sync Point		⌘,
Quantize to Grid		⌘0

Fig. 3.32 – Pro Tools LE Region Menu.

Fig. 3.33 – A 16-bar MIDI Region.

Select the Region and choose 'Loop. . .' from the Region menu. The Region looping dialog will let you repeat the region however many times you like, or you can repeat to fill a specific number of bars or until the end of the Session or next Region.

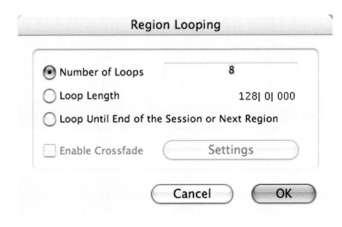

Region Looping

- ● Number of Loops   8
- ○ Loop Length   128| 0| 000
- ○ Loop Until End of the Session or Next Region

☐ Enable Crossfade   Settings

Cancel   OK

Fig. 3.34 – Region Looping dialog.

In this example I chose to loop (i.e. repeat) the region 8 times to fill 128 bars. Each repeated 16 bars is a copy of the original 16 bars and these are identified as looped regions by a small arrow turning back on itself, located in the bottom right-hand corner of the region, referred to as the Loop Icon.

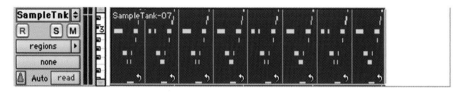

Fig. 3.35 – 16-bar region looped to fill 128 bars.

If you need to edit the loop, you can use the Unloop command to switch the 128-bar looped region back to the original unlooped 16-bar region, make your edits, then remake the long loop.

Fig. 3.36 – Unloop Regions dialog.

You also get the option to 'Flatten' the loop. In this case the looped regions are changed into normal regions that can each be edited individually without affecting the others – and the Loop icon is removed.

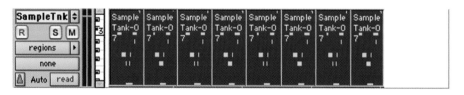

Fig. 3.37 – 'Flattened' Loops.

You may be wondering what the difference is between this way of repeating regions and the traditional way using the Edit menu's Repeat and Duplicate commands. The short answer is that Looping does gives you some extra flexibility – and those of you who have worked with Cubase or Logic may feel more at home with this way of working.

Event Menu

The Event menu gathers together the Time, Tempo, and MIDI operations windows under the first three sub-menus. The Time Operations window lets you define meter, click, and song start options by choosing commands from the popup menu at the top of the window to Change Meter, Insert Time, Cut Time, or Move Song Start. Similarly, the Tempo Operations window lets you define tempo events over a range of time and the MIDI Operations window provides several commands for changing MIDI data over a range of time.

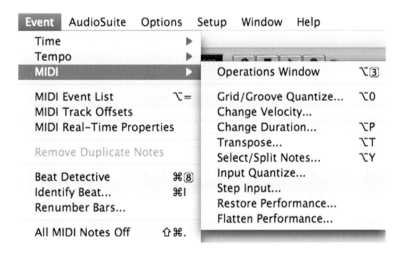

Fig. 3.38 – Pro Tools LE Event Menu.

Other Event menu commands let you open the MIDI Event List and access the Track Offsets and Real-Time Properties windows. You can also access the Identify Beat and Renumber Bars dialogs, and Beat Detective. The last item lets you issue the All MIDI Notes Off command that stops stuck notes from sounding.

AudioSuite Menu

The AudioSuite menu lets you access Digidesign's standard non-RTAS plug-ins, including D-Verb, various delays, Eqs, and dynamics processors.

When you apply any of these to an audio selection, the audio is processed to produce a new file on disk with the effect applied to it:

1. Select some audio in the Edit window and choose a plug-in from the AudioSuite menu.

2. While you are making your settings you can click on the 'preview' button to audition the effect and click the bypass button on and off to A/B the results.

3. When you are happy with the settings, click the 'process' button to create a new processed file.

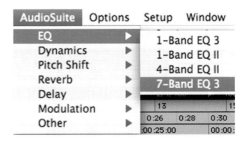

Fig. 3.39 – Pro Tools LE AudioSuite Menu.

tip ▷

If you select 'Use in Playlist', the processed file will replace the unprocessed file on the timeline in the Edit window.

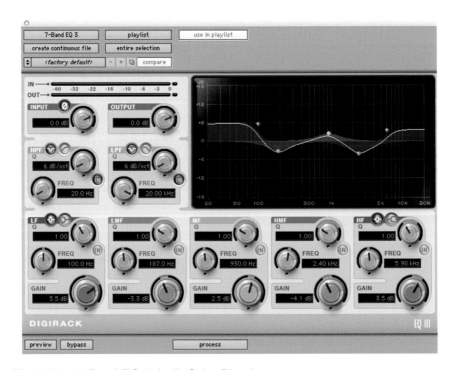

Fig. 3.40 – 7-Band EQ 3 AudioSuite Plug-in.

Options Menu

The Options menu is concerned with operational aspects. So, for example, you can choose the Record mode (Destructive, Loop, or QuickPunch), set the Scrolling options, enable Pre/Post-Roll, Loop Playback, and so forth.

Options	Setup	Window	Help
Destructive Record			
Loop Record			⌥L
QuickPunch			⇧⌘P
Transport Online			⌘J
✓ Video Track Online			⇧⌘J
QuickTime DV Out FireWire			
Pre/Post-Roll			⌘K
Loop Playback			⇧⌘L
Scrolling			▶
✓ Link Timeline and Edit Selection			⇧/
Link Track and Edit Selection			
Mirror MIDI Editing			
✓ Automation Follows Edit			
✓ Click			
✓ MIDI Thru			
Auto-Spot Regions			⌘P
✓ Pre-Fader Metering			
Low Latency Monitoring			

Mirror MIDI Editing deserves a special mention. If you edit a region while this option is enabled, your edits will automatically be applied to any and every other MIDI regions with the same name.

Fig. 3.41 – Pro Tools LE Options Menu.

tip ▷

Take the flattened loops in Fig. 3.37, for example. If you edit a note in any one of these 16-bar regions while Mirror MIDI Editing is enabled, the same edited note will appear in all the others. And if you re-select all these regions after making the edit you can turn these back into looped regions using the Region menu's Loop command – typing 1 for the number of loops.

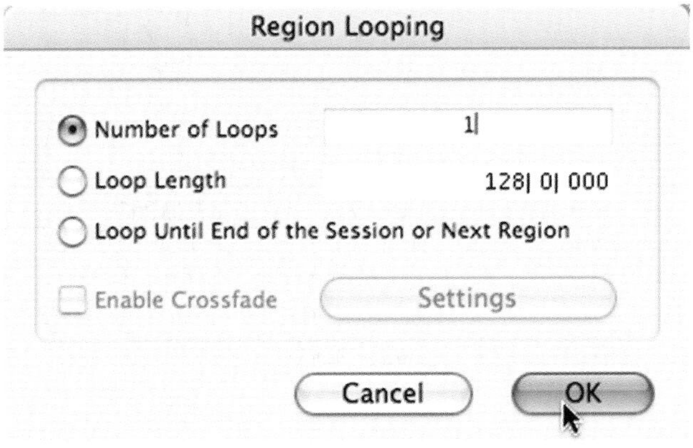

Region Looping

● Number of Loops   1|

○ Loop Length   128| 0| 000

○ Loop Until End of the Session or Next Region

☐ Enable Crossfade   (Settings)

(Cancel)   (OK)

Fig. 3.42 – Re-looping the regions.

Setup Menu

The Setup menu lets you access the Hardware Setup, Playback Engine, Disk Allocation, Peripherals, I/O Setup, and Preferences dialogs.

You can use the Hardware Setup dialog to configure your hardware interface or interfaces then use the Playback Engine dialog to tweak the settings for your computer system.

The Disk Allocation dialog lets you select folders on your disk drives to which individual tracks will be recorded.

The Peripherals dialog has 'tabs' along the top that let you select from four pages. The Synchronization and Machine Control settings are located here, along with settings for MIDI Control surfaces such as the Command|8 and Ethernet Control surfaces such as the Control 24.

The I/O Setup lets you configure the inputs and outputs, inserts and buses on your interfaces and rename these to reflect their usage if you wish.

QuickTime Movie Offsets can be set and you can access the Session setup window and various MIDI setup dialogs, the Click and the Preferences dialogs.

If you have DV Toolkit installed, three additional items appear in the Setup menu that you can use with time code when working to picture (See Appendix 2).

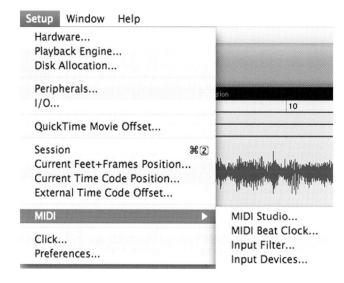

Fig. 3.43 –
Pro Tools LE
Setup Menu.

Window Menu

The Window menu lets you show or hide the different windows, as necessary. Most of these can be opened using keyboard commands that you will get used to over time as you work with the software.

The Undo History gets a special mention at this point. This is a 'life-saver' for anyone learning to use Pro Tools LE – or even expert users who change their minds about an earlier edit for creative reasons. The Undo History window lets you view a list of undoable and redoable operations since you opened the Session. It can even display the time of day that you made each edit, so you can always take your session back to the state it was in at a particular time.

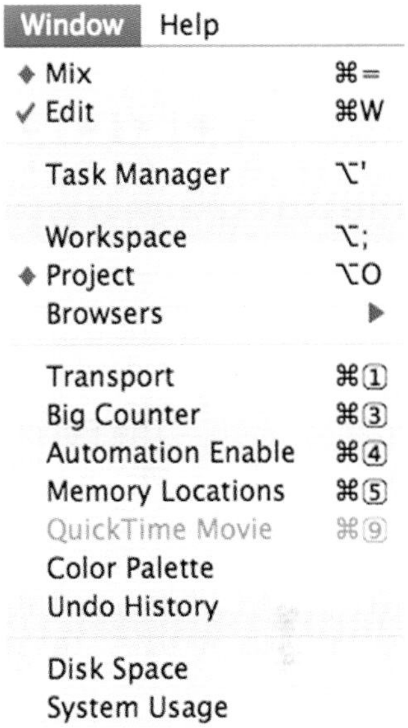

Fig. 3.44 – Pro Tools LE Window Menu.

Help Menu

Finally, there is a Help menu. This contains links to the most often-used documentation in Acrobat format, such as the Reference Guide and the Plug-ins Guide.

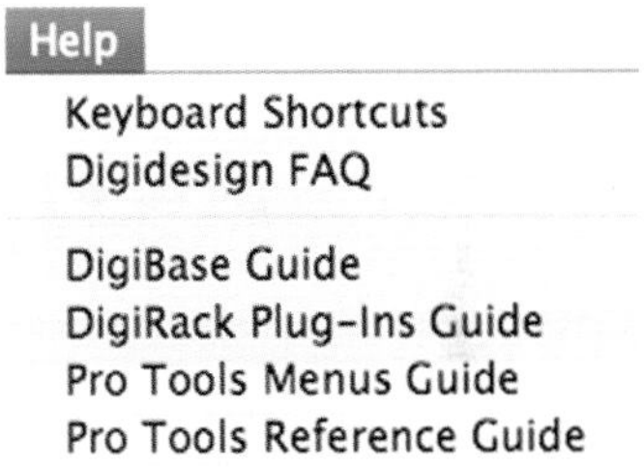

Fig. 3.45 – Pro Tools LE Help Menu.

Browser, browser, browser. . .

As with any software, when you work with Pro Tools LE regularly you are going to generate a lot of files on your disk drives. To help you keep track of all these and find your way around your disk drives, including any networked drives, Pro Tools LE has three types of so-called DigiBase browsers – the Workspace browser, the Volume browser, and the Project browser. These browsers are essentially databases that can be used for searching, sorting, auditioning and importing audio, MIDI, and session files.

The Workspace browser provides access to all your mounted disk volumes, as well as the folders and files they contain. The Workspace browser lets you work with files similarly to the way you would in the Macintosh Finder – finding, copying and deleting files and creating folders.

If you double-click on a volume in the Workspace browser, a Volume browser opens in a new window. The Volume browser lets you manage the files on your local and any networked volumes, allowing you to conveniently view, audition and import individual items.

note ▷ Volumes are partitions on hard disk drives, network storage devices, or CDROMs. Like the volumes of a book, hard drives can be arranged as one or more volumes or sections. So, just as you might browse through a book volume, you can use Pro Tools LE's Workspace, Project and Volume Browsers to browse through your disk volumes to see what you have stored there.

Fig. 3.46 – Workspace Browser with the Audio drive's Volume Browser open in front of it showing the files and folders on disk.

The Project Browser shows all the files being used in your current session. This is particularly useful when you have a large project with lots of files – some of which may be stored on disk drives other than the drive where your project is stored. You can use this browser to check where the audio files are actually stored on disk, how big they are, what duration they are, when they were created and so forth.

tip ▷ You can audition files in any browser any time you like by simply clicking on the small speaker icon next to the small waveform display.

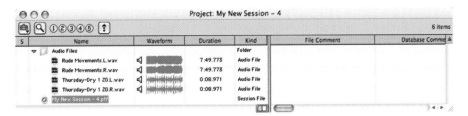

Fig. 3.47 – The Project Browser showing a Session file and its accompanying Audio Files.

You can drag and drop items from these browsers directly onto the Timeline or Regions List of your current Pro Tools session. You can also use these browsers to manage various Pro Tools tasks that can be carried out in the background – or to find missing files.

The Task Window

The Task window lets you view and manage all the tasks that Pro Tools LE frequently has to carry out such as copying and converting files, searching for files, indexing files and creating fades. These tasks occur in the background, and you can continue to record, edit and mix while these are being taken care of 'behind the scenes'. You can use the Task window to check on progress, pause, or cancel background these tasks as necessary. To open the Task Window, choose Task Manager from the Window menu.

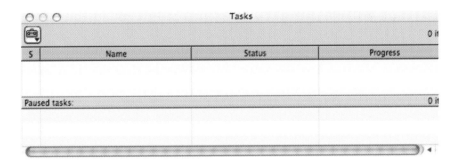

Fig. 3.48 – The Task Window.

The Relink Window

Sometimes, a Pro Tools LE Session file loses track of its associated media files. This can happen if you re-configure your computer and its data storage systems, for example.

When you next open the Session, if it cannot find all the files it expects, you are presented with an option to 'Manually Find and Relink' the missing files. If you choose this option, the Relink window will open to let you relink the Session to the media files.

If the Session is already open and you know that there are missing files that need to be relinked you can access the Relink window from the Project window.

Fig. 3.49 – Project Browser with the local Browser menu open to reveal the Relink Selected command.

Open the Project window from the Window menu, select an audio file that needs relinking, then choose Relink Selected from the Browser menu to open the Relink window.

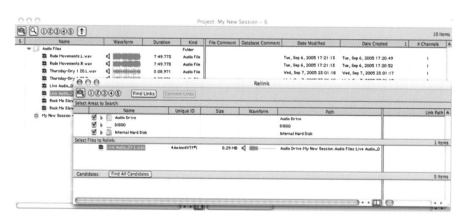

Fig. 3.50 – The Relink Window.

You can pretty much follow the instructions in this window to relink files. Typically, you select the files to relink, press a button to 'Find All Candidates' that may be the missing files, wait for Pro Tools to find the files, then approve which of the found files (if any) to relink. A separate DigiBase Guide PDF file is supplied with Pro Tools LE that explains all about the DigiBase browsers.

Drag and Drop Features

Pro Tools lets you drag and drop audio, MIDI, region group, REX, and ACID files from the desktop to the Timeline or Region List. You can also drag and drop a Pro Tools session file to the Timeline to open the Import Session Data dialog.

If you select one or more regions from the Regions List and drag these to the Timeline, Pro Tools will either place them into a single track or each region into its own new track, depending on the Timeline Drop Order that you choose.

At the bottom of the Region List, the Timeline Drop Order submenu offers two choices: 'Top to Bottom' and 'Left to Right'.

When Top to Bottom is enabled, regions spread top to bottom, creating new tracks for each region.

When Left to Right is enabled, regions spread across the single destination (drop) track or a newly created track.

Audio files can also be dragged from any DigiBase browser to a plug-in window to quickly load samples into plug-ins, such as Digidesign Synchronic.

You can also drag and drop MIDI data from a plug-in to the Timeline, Track List, or to the Region List – if the plug-in supports this feature.

What this chapter has covered

You should now have a basic understanding of what the Pro Tools LE software looks like and what it can do in terms of how many tracks it has and so forth.

You should have familiarized yourself with the three main windows (Edit, Mix, and Transport) and with the main features of these windows.

You should know which modes and which tools are available and how to find your way around in the Edit window.

After reading through the menus section you should have a reasonable idea of which menu to go to if you want to change the view or to add a new track or whatever.

You should understand what the browsers are and how to use the drag and drop features.

Don't expect that you will remember all this stuff straight away, especially if you are new to Pro Tools – or even if you are upgrading from a previous version. It will always depend on how much time you spend working with the software and how motivated you are to get a handle on all this stuff.

There are so many possible configurations of Pro Tools LE systems that it is not practical to give you a comprehensive set of step-by-step instructions for all these. Fortunately, the instructions given in the Pro Tools LE manuals are generally very clear and easy to understand. In truth, setting up is easy enough. Connect the audio interface to your computer using the FireWire or USB cable that comes with the system. Hook up the microphones, instruments, a pair of powered speakers and any other equipment that you intend to use. Then load the Pro Tools LE software onto your computer, open a new (or existing) session – and you're away!

tip ▶ Digidesign supplies Demo Sessions on the software installation discs that you can use to help test your system to make sure it is working properly. It is always a good idea to open one of these demo sessions, hit the Spacebar on your computer keyboard, and make sure that it is playing back OK if you are in any doubt about whether everything is hooked up correctly or not.

After you have installed your hardware and software and made sure that it is working there are still several things you need to know about before you start using your Pro Tools LE system in earnest.

This chapter will help you to get your LE system set up and ready for action, taking you through the basics of opening and saving your first Session, choosing sample rate and bit depth, allocating hard disk space, adding tracks to work with, and so forth.

Often, you will want to import audio tracks, or transfer audio from CD or other sources. And if you are working to picture you may want to import a QuickTime movie or the audio from that movie. All will be revealed. . .

Installing the software

You should always check which version of the system software is installed on your computer before installing application software. You can do this by choosing 'About This Mac' from the Finder's Apple Menu and reading the version number there.

Fig. 4.1 – About this Mac window.

If you are using Windows, the System Properties Control Panel shows the Operating System version number in a similar way. At the time of writing, for example, Windows XP with Service Pack 2 was required.

The Digidesign website (www.digidesign.com) has a Compatibility section that you can access from the Support page. This always lists the versions of your computer's operating system that will work with the version of the Pro Tools software that you have.

After installing the software that comes on the discs supplied with your hardware, you should check Digidesign's Compatibility page to see if there are any recommended operating system or application software updates.

note ▷ Never be tempted to automatically update your computer's system software – always check the Digidesign website first. It can be time consuming and awkward to have to re-install earlier versions of your computer's system software if you inadvertently or unwisely update the system software to an incompatible version and thereby prevent your Pro Tools application software from working correctly.

tip ▷ After installing the additional software that comes bundled with your system, check the websites for updates. And consider upgrading Reason Adapted, Ableton Live, the IK Multimedia and other bundled software to the full versions.

Restoring the default Preferences

Before following the examples through in this chapter you will need to restore the default Preferences in Pro Tools LE. As you make changes to your Pro Tools LE software configuration, which can happen even when you run the Demo Session that comes with Pro Tools LE, these changes are stored in a Preferences file and applied the next time you open a new Session.

First you should Quit Pro Tools LE if it is running. Then, on Mac OSX, go to the Users folder in your boot drive and double-click your Home icon to reveal your personal folders. Open the Library folder and you will find the Preferences folder within. Look for a file named 'Pro Tools LE v7.0 Preferences', drag this to the Trash and empty the trash. Then restart your computer, launch Pro Tools LE and the screens will look similar to those in this chapter.

If you are using Windows XP, look in your boot drive (typically the C drive) for the Program Files folder. Look inside this for the Common Files folder. Inside this you should find a Digidesign folder containing a DAE folder that contains the DAE Prefs file. Drag this to the Recycle Bin. Go back to the top level of the boot drive (usually the C drive) and open the Documents & Settings folder then open the current User Folder. Here you should see the Application Data folder.

tip ▷ If you cannot see the Application Data folder it must be hidden. You can make it visible from the Tools Menu. Open Folder Options: View:Hidden files and folders. Select Show hidden files and folders and click Apply.

Inside the Application Data folder there will be a Digidesign folder containing the Pro Tools Preferences file. Drag this to the Recycle Bin as well, and empty this. After trashing the preferences, restart your computer, launch Pro Tools LE and the screens should look similar to those in this chapter.

Opening a New Session

Launch Pro Tools, choose New Session from the File Menu or use the Command-N keyboard command (Control-N on the PC), then select the Audio File Type, Sample Rate, Bit Depth and I/O Settings for your session.

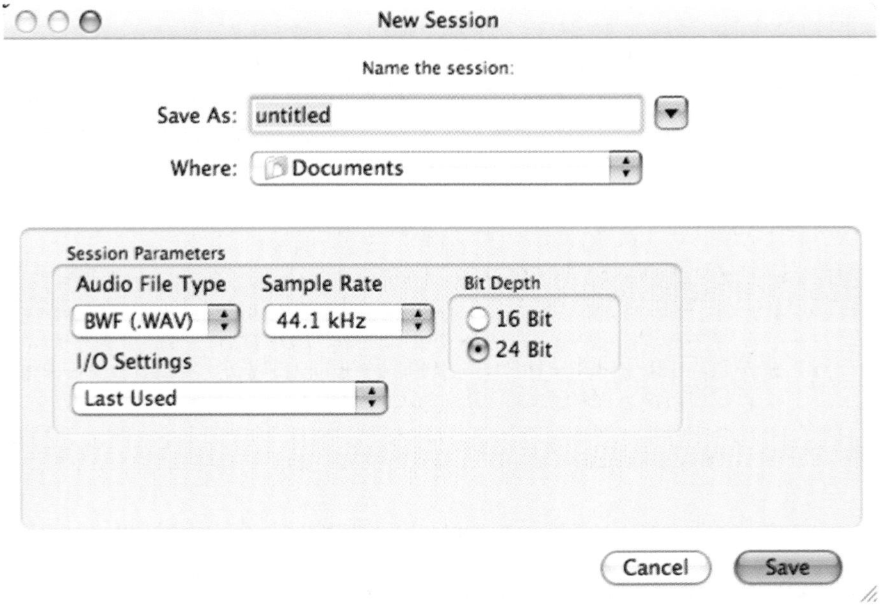

Fig. 4.2 – New Session dialog.

Choosing File Types

The best choice of file type for compatibility today is .WAV. Older Pro Tools systems used Sound Designer II (SDII) format and the Audio Interchange File Format (AIFF) used to be popular when moving files between different computer platforms.

Choosing Sample Rates and Bit Depths

When analogue audio is converted to digital audio in the analogue to digital converter (A–D) converters in your Pro Tools interface, the audio waveform is sampled many times each second.

There are four different sample rates that you may encounter with Pro Tools LE systems – 44.1 kHz, 48 kHz, 88.2 kHz and 96 kHz. The first of these, 44.1 kHz is the sample rate used by CDs and other popular digital audio equipment. The second, 48 kHz, is the sample rate used by DAT and other digital audio equipment – especially when associated with video equipment. The higher

88.2 and 96 kHz rates (supported by Digi 002 systems, e.g.) can be used on DVD-Audio or DVD-Video discs or with other professional audio equipment.

tip ▷ Always use 44.1 kHz sampling rate unless you have a specific reason to use 48 kHz or higher. This saves you having to convert to 44.1 kHz to burn onto CD and keeps your file sizes more reasonable.

note ▷ If you double the sample rate you double the file size. So 1 minute of mono audio recorded at 44.1 kHz/16 bit would use about 5 megabytes of disk space and at 96 kHz/16-bit it would occupy 10 megabytes. And if recorded at 96 kHz/24-bit it would occupy 15 megabytes of disk space.

What kind of 'sampling' is going on here? Well, what actually happens in an analogue to digital (A–D) converter (like those in your Digidesign audio interface) is that the voltage value of the waveform at each sample point in time is measured and a numerical value corresponding to this voltage is output from the converter as a binary number. So if you sample the analogue audio waveform 44,100 times each second, then the A–D converter spits out 44,100 16-bit binary numbers each second – and these are what get recorded onto your hard disk drive. So digital audio is just a bunch of 1s and 0s.

The bit depth (or resolution) of the converter is the number of bits, that is, binary digits, (in other words the 1s and 0s) used to represent the value of the sample. The more bits available to represent this number, the finer the measurement of the voltage value of the waveform, and, consequently, the more accurate the representation of that waveform.

note ▷ Having more bits available to measure with is like having more sub-divisions on your ruler or tape-measure: you can measure to smaller fractions of an inch or to hundredths or thousandths of a metre depending on the available sub-divisions.

These binary numbers are stored in a computer file on your hard disk. When you play this audio back, the series of bits is presented to a digital to analogue (D–A) converter that rebuilds an analogue waveform that corresponds as closely as possible to the original.

Basically, the more samples of the audio waveform that you take each second and the more bits you use to represent the values of these waveform samples, the higher the quality of the sound that you will store and subsequently reproduce.

tip ▷ As far as Bit Depth is concerned, although 24-bit is recommended for professional use, 16-bit will be fine for many projects – such as demos.

note ▷ If you record at 24-bit, using 8 more bits of information than 16-bit, your audio files are half as large again compared with 16-bit files. For example, 1 minute of mono audio, which uses about 5 megabytes of hard disk space at 44.1 kHz/16-bit, would use about 7.5 megabytes at 44.1 kHz/24-bit.

Saving your Session

Having chosen your file type, sample rate and bit-depth, the next step is to choose the disk drive and folder where you want to save your Session – before you hit the Save button.

If you don't see the expanded New Session dialog, just click the small arrow to the right of the 'Save As' text field to expand it so that you can see your disk drives and folder structures.

Fig. 4.3 – Expanded New Session dialog.

note ▷ The New Session dialog defaults to showing the Documents folder in your User folder until you change it. The next time you choose Save As, it will show the last directory that you saved into. If you hit Save without selecting where to save your Session, it will save into whichever directory was selected previously. So if you can't find it, now you should have a better idea about where to look!

tip ▷ If you are not quite sure where you want to put your files or how to get there, save the Session onto the Desktop first – where it is easy to see it and find it – and you can always move it to a better place later.

The biggest problem for new users is to know where they have saved their files. Keep in mind that it is always your responsibility as the User to decide where you want to save your files, but it is just too easy to forget to do this and then be left wondering where the file went! If you can remember the name you gave the file, you can always use your computer's 'find file' function – but this won't work if you have forgotten the filename or if you spell it wrongly – or don't even know how to use the computer to find a file! So the best thing to do is to get it right in the first place and save to a folder of *your* choosing.

Using the Expanded New Session dialog, you can move the scroll bars up and down or to left and right to reveal the different disk volumes and directories (folders) available on your system. An initial investment in time spent getting familiar with how this works will repay you handsomely in the future in frustrations avoided – frustrations that will certainly occur if you remain unsure about how all this works.

note ▷ The file system on Windows computers is very similar, and the same comments about your personal responsibility to gain familiarity with this apply equally.

What gets saved

Now let's take a look at what gets saved onto your hard drive. Use the Hide Pro Tools LE command from the Pro Tools menu if you are working on the Mac (click the Minimize button if you are working on the PC) and take a look at the hard drive where you saved your Pro Tools LE Session. Or, better still if this is the first time, simply close your Pro Tools LE Session for the moment.

If you saved onto your desktop, you will see the folder immediately. Double-click on this to look inside.

The most important file right now is the Session file with its '.ptf' extension. This is the document that displays the Edit and Mix windows and keeps track

Fig. 4.4 – New Session folder.

of the Audio Files, Fade Files, Region Groups and other elements used in your Session.

You should also see at least three empty folders – Audio Files, Fade Files, Region Groups – along with a Session File Backups folder containing however many backup files you have elected to use in the AutoSave Preferences.

As the names suggest, as soon as they are created, audio files are stored inside the Audio Files folder, fade files in the Fade Files folder and so on.

tip ▷ If you put your new Session folder onto your desktop initially, now would be a good time to move it onto your dedicated audio drive. And don't forget to trash the folder from your desktop before carrying on with your project – to prevent confusion.

Why you should use an external hard drive

It is always best to use a fast external hard drive with a rotational speed of 7200 rpm or more to record your Pro Tools LE sessions onto. There are several reasons for this. Firstly, the internal drives on laptops and on many desktops are slower hard drives with rotational speeds of 5400 rpm or less, so you won't be able to record as many tracks or use as many edits per track with these. And if you are doing a lot of recording you will need more disk space than is available on your internal drive anyway. Then, especially when you have made lots of edits, a time will come when the hard disk becomes fragmented and begins to slow down. If your Pro Tools files are all kept on an external drive, it is much easier to get rid of the disk fragmentation than if they are on the internal drive with your operating system software.

tip ▷ The best fix for disk fragmentation is to back up all your files to DVD-R or CD-R, erase the disk completely, then copy back onto the hard disk any projects that you need to work on further. You can use software such as Norton Utilities to defragment the files while they are still on the disk, but erasing the disk is a surefire method of removing fragmented files that costs nothing – and any excuse to make sure you backed up your files has to be a good thing!

note ▷ If you are using an ATA/IDE or FireWire hard drive, you should initialize (erase) your drive with Windows Disk Management if you are using a PC or with the Disk Utility application included with Apple System software if you are using a Macintosh.

You can install additional internal drives on most desktop computers to serve as dedicated audio drives for Pro Tools LE. However, nothing beats the convenience of an external drive if you want to take your projects to your friend's house to work on his or her system or to a professional studio to mix or master your music there.

Be aware that some hard drives may not give good performance (or may not even work at all) with Pro Tools systems. To check which hard drives are recommended for use with Pro Tools LE, visit the Digidesign Website and look at the section on compatible products.

Allocating Hard Disk Space

When you create a new Session in Pro Tools, audio files will normally be recorded into the Audio Files folder inside the Session folder that you have newly created.

note ▷ If you save to the desktop, this means that you are recording to the Internal hard disk (or whichever hard disk drive contains the operating system).

Select Disk Allocation from the Pro Tools Setup menu and take a look. In the example given earlier where the new Session was saved onto the Desktop, the Disk Allocation dialog shows that the Audio tracks will be recorded into the folder named My New Session on the disk drive named Internal Hard Disk.

Disk Allocation		
Track	Root Media Folder	
Audio 1	Internal Hard Disk:My New Session:	‡
Audio 2	Internal Hard Disk:My New Session:	‡
Audio 3	Internal Hard Disk:My New Session:	‡
Audio 4	Internal Hard Disk:My New Session:	‡

Fig. 4.5 – Disk allocation onto the Internal Hard Disk.

If you then close this session and move the session folder onto an external audio drive, re-open it from the audio drive and start recording audio, the Disk Allocation will *still* be set to the Internal Hard Disk. So Pro Tools will create a new folder with the same name as your session (My New Session in this example) on the hard drive originally allocated (the Internal Hard Drive in this example) and will record the audio into a new Audio Files folder inside that session folder.

If you do not realize that this has happened, things could get quite confusing for you! The Pro Tools Session on the Audio drive will still be able to access any audio files recorded onto other drives in your computer system, but a big problem arises if you take your audio drive (or a straight copy of the Session on CD-R or whatever) to another computer. In this case, the Session cannot access any audio files inadvertently recorded to the internal hard drive on your home computer. After all, this may be hundreds of miles away – and even if it's only across town it could be a major hassle!

Fig. 4.6 – Duplicate Session folder created on Internal Hard Disk containing newly recorded audio files when Disk Allocation is not set correctly.

The best way to deal with this situation is to avoid it happening in the first place by checking the Disk Allocation before you start recording any new audio.

Each audio track has a 'Root Media Folder' listed into which the audio for that track will be recorded. Click on any of the listed folders and you will see a popup selector that lets you choose which hard drive partition to record onto.

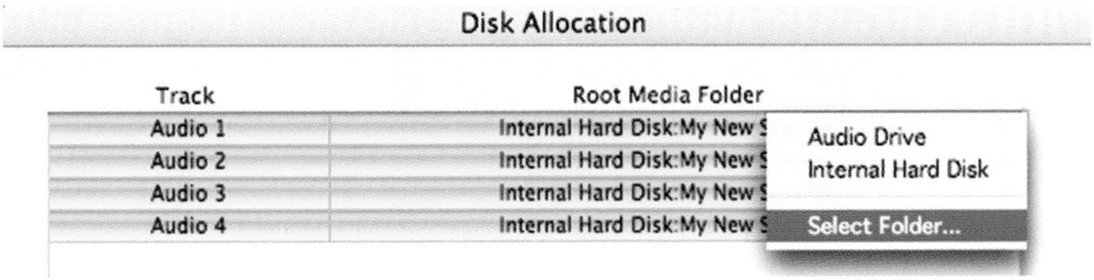

Fig. 4.7 – Root Media Folder popup selector in the Disk Allocation dialog.

tip ▷ To allocate all the audio tracks to record into your new Session folder, hold the 'Alt' key on your computer keyboard, choose the Select Folder option from any of the Root Media Folder popups, navigate to the Audio Files folder in your new Session folder on your audio drive and choose this.

Fig. 4.8 – Choosing a Disk Allocation folder.

Any further audio tracks that you create should now default to using the correct audio files folder in your session folder on the correct hard drive – in this case, the one named Audio Drive.

Fig. 4.9 – Disk Allocation correctly set.

Round Robin Disk Allocation

You may be wondering about the option to 'Use round robin allocation for new tracks'. If this option is checked, Pro Tools automatically distributes the disk allocations for any newly created tracks among the drives connected to your system – apart from the startup drive. The idea here is to make more efficient use of the available bandwidth if you have several audio drives available for use.

If this is the case, Pro Tools creates an additional Session folder on each additional audio drive and records each alternate audio track into the next available drive.

note ▶ You can exclude from this round robin allocation any other drives connected to your system by making these volumes safe – designating them as P (Playback only) or T (Transfer) in the Workspace browser.

The problem that can arise if you use round robin allocation is similar to the problem mentioned earlier. It can be all too easy to forget that the audio files for this Session are scattered around these different hard drives when copying, moving or backing up your Sessions if you have used round robin allocation.

And if the previous person who used your system left this option checked and you don't uncheck this before recording new audio, you may not even be aware that this is happening. The golden rule here is to always make sure that this option is not checked before you do any audio recording – unless you really do want to use it.

note ▷ It's unlikely that the Round Robin Allocation option would be needed for anything other than the largest and most complex sessions in Pro Tools LE.

tip ▷ Now despite all my warnings here, some of you are going to fall into one or other of these traps. But there is one more thing you can do that will help to prevent you from losing track of your audio files: when you make copies of your Sessions, especially when you intend to move these onto other drives to make backups, or when you are preparing to take your sessions to other studios, always use the 'Save Copy In. . .' command from the File menu and check the 'Items To Copy: All Audio Files' option. This creates a copy of your Session folder complete with all the audio files from whichever drive you have recorded these onto.

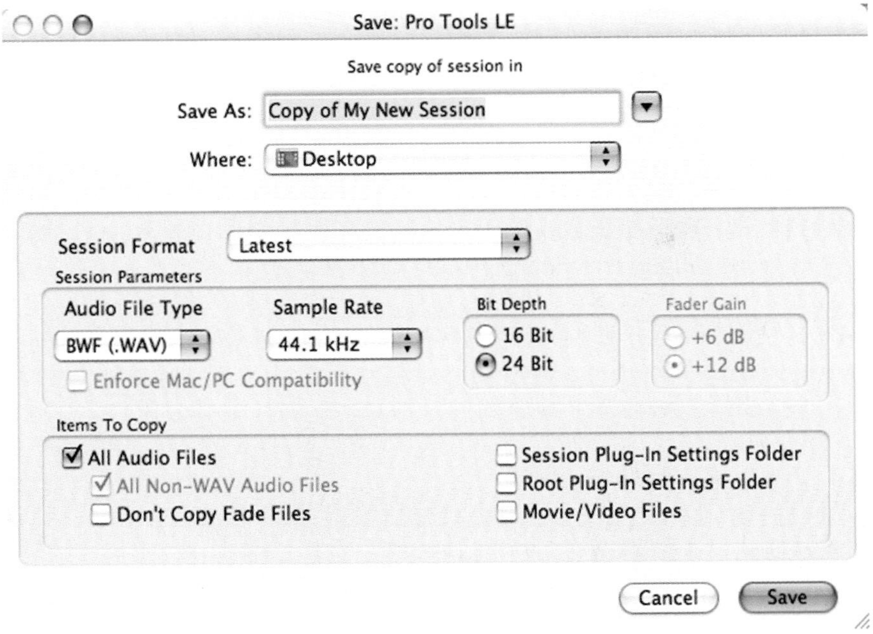

Fig. 4.10 – Save Copy In. . . dialog.

Record Allocation

While we are on the subject of hard disks, there is just one more related topic worth mentioning here.

Pro Tools defaults to providing an open ended record allocation. This means that Pro Tools lets you record as much audio as there is free space for on your hard disk.

If you leave this option open ended, Pro Tools has to check the hard disk to see how much space is available before going into Record, which may lead to a short, but annoying, delay after you hit Record.

If you open the Operation page in the Preferences window you will see a section that lets you change this. I recommend that you limit the record allocation to just a little longer than the longest time you expect to record for.

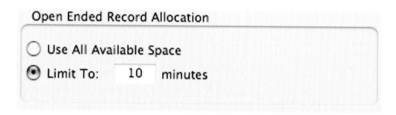

Fig. 4.11 – Changing the record allocation in the Operation Preferences window.

Setting up the inputs and outputs

If you are just getting started with a new Session, or if you have opened a Session created on another system, it is worth setting up the Inputs and Outputs in the I/O Setup window that you can open from the Setup menu.

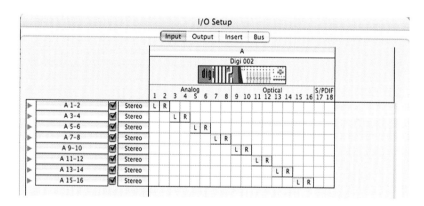

Fig. 4.12 – I/O Setup window.

If you are using a Digi 002, for example, just hit the Default button and the default names of the Inputs will appear.

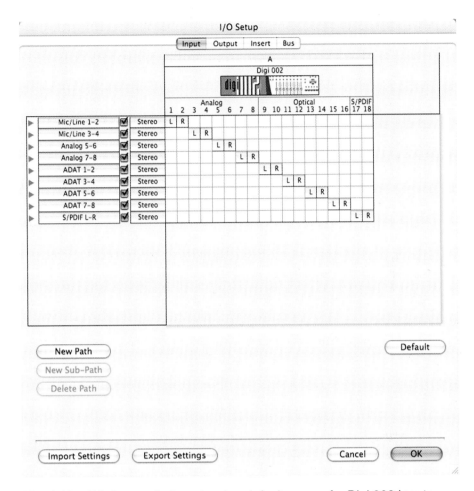

Fig. 4.13 – I/O Setup window showing default names for Digi 002 Inputs.

Switch to the Output and Insert pages using the tabs at the top of the window and do the same for these.

tip ▶ If you always have a particular device, such as a Shure SM57 or SM58 microphone, connected to a particular input, you can type these names into the name fields instead. Just double-click on an Input name, type what you like, and hit Return.

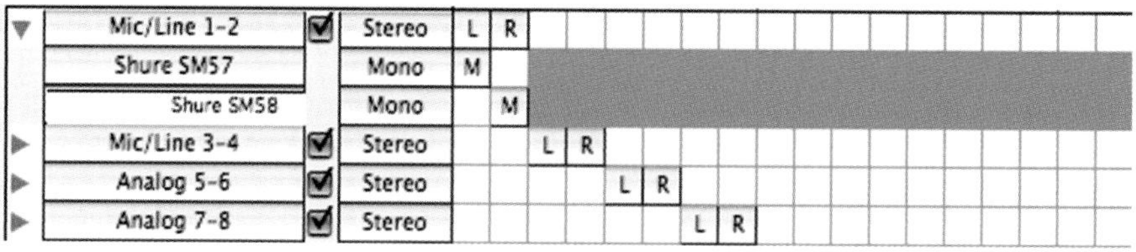

Fig. 4.14 – Part of the I/O Setup window showing individual names for inputs.

You can do the same thing with the Outputs, Inserts and Busses. This can be very helpful if you always use particular busses to route audio to a reverb plug-in on an Auxiliary channel, for example. Selecting Reverb instead of Bus 1–2 is much easier once you have this set up.

Having taken the trouble to do all this, you can save your I/O settings for use next time you open a new Session or bring in a Session created on someone else's system. Just hit the Export Settings button in the I/O Setup window and type a suitable name for your settings file. By default, Pro Tools will save this into the I/O Settings folder inside your Pro Tools LE Folder – which is normally installed inside the Applications folder in your 'root' hard disk (the 'boot' disk containing the operating system).

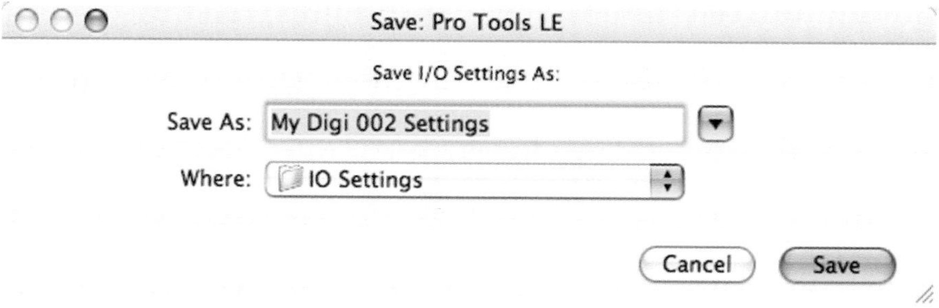

Fig. 4.15 – Saving Pro Tools LE I/O Settings.

Now, when you open a New Session, you can choose these settings from the popup selector in the New Session dialog box.

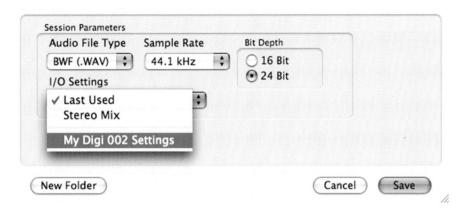

Fig. 4.16 – Choosing I/O Settings in the New Session dialog.

Setting up your first new session with Audio and MIDI tracks, Auxiliary inputs and Master faders

When your new session first opens, it has no tracks and it is not set up properly – it is like a 'blank palette' waiting to be sorted out.

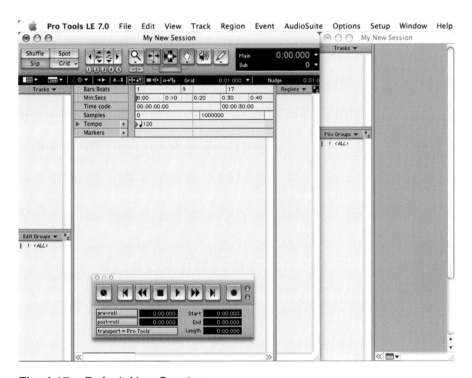

Fig. 4.17 – Default New Session.

The first thing you should do is to tidy up your windows. Click the double arrow icons near the bottom left and bottom right of the Edit window to close up the Show/Hide Tracks and Edit Groups lists at the left and the Audio and MIDI Regions lists at the right. Do the same thing with the Mix window's Tracks and Mix Groups lists. Then you should decide which rulers, if any, you need to display at the top of the Edit window. If you are working on a music project you will probably want to leave the Bars:Beats and the Min:Secs rulers visible along with the Tempo and Markers rulers. You may also wish to open the Meter ruler.

This is probably a good time to set various preferences as well, such as the AutoSave. You will find this in the Operation panel in the Preferences window that you can access from the Setup menu (or from the Pro Tools LE 7 menu on the Mac).

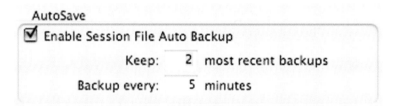

Fig. 4.18 – AutoSave Preferences.

note ▷ By default, the 'Timeline Insertion Follows Playback' preference in the Operations menu is off. So whenever you stop playback during your Pro Tools Session, the counter returns to wherever it was when you started playback. So, for example, if you start playback from Bar 1:Beat 1 the counter will go back to that location when you stop, or if you start playback from Bar 6:Beat 3 it will go back there when you stop.

If you prefer that the counter stays wherever it is when you stop playback, you can tick the box next to 'Timeline Insertion Follows Playback' in the Operation Preferences page to change the playback mode.

tip ▷ You can switch the 'Timeline Insertion Follows Playback' preference on and off by pressing the 'N' key on your computer keyboard. And be aware that it is all too easy to do this accidentally – a gentle touch on the key is all it takes.

I usually make sure that the Transport window is fully expanded to show all the MIDI controls and counters by selecting these options from the View menu. I also set the Main counter to either Bars:Beats for music projects or

Time Code for post-production projects, again by choosing these options from the View menu – or using the popup selectors (the small downwards pointing arrows to the right of the Main and Sub-counters). I often set the Sub-counter to Min:Secs so that I can always see the elapsed time.

Fig. 4.19 – Fully expanded Transport window showing MIDI controls and counters.

With this done, it is time to start adding tracks to build up your mixer. Choose New Tracks from the Track menu, create a couple of mono audio tracks.

Fig. 4.20 – New Tracks dialog.

Then click the + sign at the right to add a second line and create a couple of stereo tracks. Continue this way until you have added as many tracks as you think you will need for your project, typically finishing off by adding a stereo Master fader.

Fig. 4.21 – New Tracks Dialog creating multiple track types.

The Edit window should look something similar to the accompanying screen-shot in which the track heights are set to 'Small' by clicking on the vertical ruler area to the right of each track.

tip ▷ Hold the 'Alt' key while you select track height to change all the track heights to the same setting.

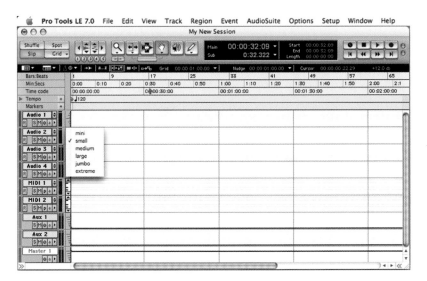

Fig. 4.22 – Pro Tools LE 7 menu bar at the top with the Edit window below this showing the track height selection.

The Mix window should look something like the next screenshot, depending on which items you have chosen to view using the View menu.

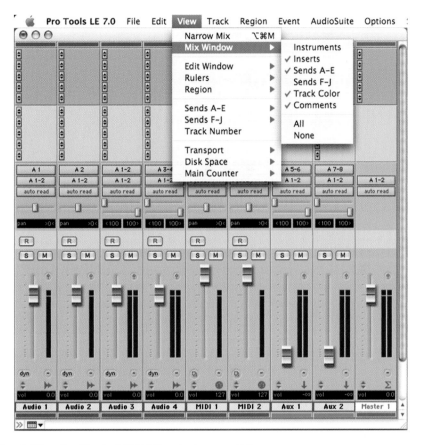

Fig. 4.23 – Pro Tools LE 7 Mix window and View menu.

note ▷ Audio tracks, Auxiliary Inputs, and Instrument tracks include one Insert View and two sets of Sends Views: Sends A–E (sends 1–5) and Sends F–J (sends 6–10). Instrument Tracks also have an Instruments View at the top of the Channel Strip. The View menu lets you choose which of these to show in the Mix and Edit windows.

I usually have the Inserts and the first set of Sends visible in the Mix window and keep these hidden in the Edit window.

Some users prefer to work from just one window, in which case it can be convenient to show the Inserts and the Sends in the Edit window along with the I/O view that provides access to inputs, outputs, volume and pan controls.

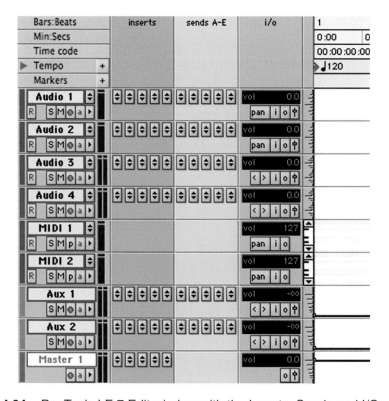

Fig. 4.24 – Pro Tools LE 7 Edit window with the Inserts, Sends and I/O views visible.

At this point you should tidy up your screen, resizing the windows and positioning them to make best use of your available screenspace.

tip ▷ It can be handy to have the Big Counter window open as part of your standard setup. This lets you see the playback position at a glance and it can also help with navigation – just type numbers directly into this window and press the Return key on your computer keyboard to jump to that position (even when already playing).

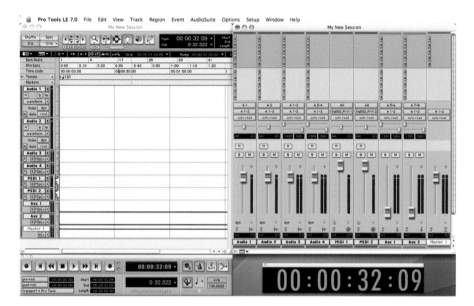

Fig. 4.25 – Pro Tools LE 7 Edit, Mix and Transport windows plus the Big Counter.

Importing MIDI and Audio

With your Pro Tools Session set up for action, you might just start recording right away. On the other hand, you may want to bring in audio or MIDI to work with from other sources.

If you have prepared MIDI files using a sequencer, you can use the 'Import MIDI to Track' or 'Import MIDI to Region List' commands in the File menu and use these as your starting point.

'Import MIDI to Track' conveniently creates a new MIDI track for each track in the MIDI file that you import.

'Import MIDI to Region List' just puts the MIDI tracks into the Region List and leaves you the task of placing these into MIDI tracks that you have to create.

Importing audio works similarly. You may want to use a track from a CD as a 'template' to work to or perhaps someone has sent you some audio on CD (or on a CD/DVD-ROM or whatever) to work with. You can use the 'Import Audio to Track' or 'Import Audio To Region List' commands to bring in audio from CD or other storage devices.

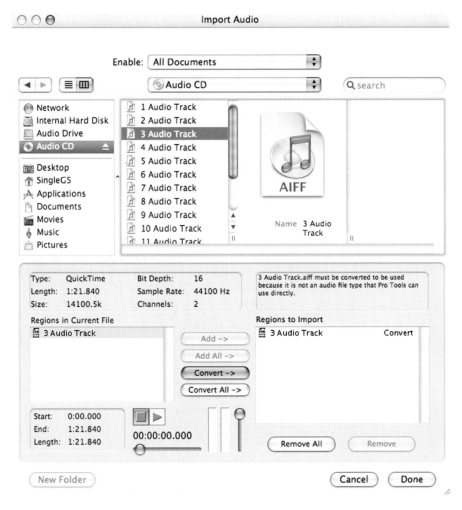

Fig. 4.26 – Import Audio Dialog.

Select the audio track or file that you want to import from the file browser area near the top of the dialog window, and any regions within the file will be displayed in the area to the left below this.

note ▶ Most audio files (and CD tracks) will just contain one region encompassing the entire track. Other audio files may contain two or more regions that have previously been defined.

Choose the region or regions that you want to import by clicking (or shift-clicking for multiple regions) in the lower left area to select these. Then click the 'Convert' or 'Convert All' buttons to add these to the 'Regions to Import' list on the lower right hand-side of the Import Audio dialog.

When you hit the 'Done' button, another dialog appears to let you choose the folder into which the imported audio will be placed after conversion to the audio format you are using for your session.

Fig. 4.27 – Choose a Folder dialog.

Finally, Pro Tools creates an audio track for you and places the imported audio file into this as well as into the Audio Regions list.

Fig. 4.28 – Audio imported to a track also shows up in the Audio Regions list.

Using the Workspace Browser to import audio using drag & drop

You may find it easier to use the Workspace Browser to import audio CD tracks directly into Pro Tools sessions. The Workspace is a file 'browser' that provides access to all your disk drives and to the folders and files within these. It is set out rather like a spreadsheet with columns that display meta-data for each item in the list. The bit-rate and sample-depth are displayed, for example, and the Workspace browser provides comprehensive search capabilities. You can also audition tracks on the CD by clicking the loudspeaker icons in the Workspace's Waveform column.

The neat thing here is that you can simply drag and drop any audio file or any CD track that is available in your Workspace onto any audio track in the Edit window of the current Pro Tools Session – or into the Audio Regions list.

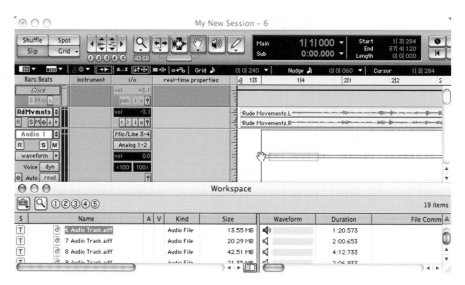

Fig. 4.29 – Dragging & Dropping a CD track from the Workspace to a Track in the Edit Window.

Importing Loops & Samples

You can also import audio (or MIDI, region group, REX, or ACID files) by dragging and dropping from the Macintosh Finder or from Windows Explorer directly into your Pro Tools Session. You can either drag and drop a file into an existing track or drag and drop the file into an empty part of the Edit window, in which case Pro Tools will create a new track automatically.

Fig. 4.30 – Dragging and dropping an audio file from the desktop into Pro Tools.

Transferring audio into Pro Tools from CD, DAT or other digital sources

Many of you will have a CD player or a DAT machine in your setup and will want to play back a tape or a CD from one of these machines and transfer this digitally in real time into Pro Tools.

Typically, you will connect the S/PDIF output from the player to the S/PDIF input on your Pro Tools interface and select this as the input to a stereo track in your Pro Tools Session.

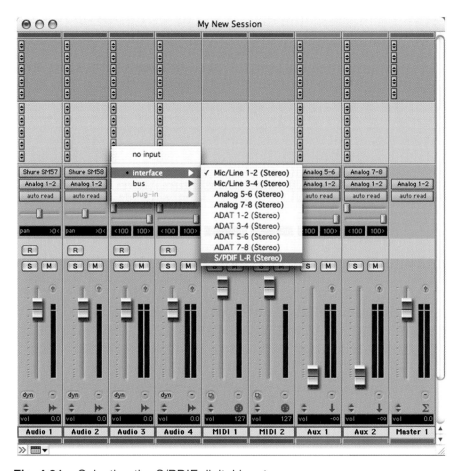

Fig. 4.31 – Selecting the S/PDIF digital input.

You should also check the settings in the Hardware Setup dialog that you access from the Setup menu.

Some CD players use Optical S/PDIF connectors, like ADAT optical connectors, so you will need to change the Digital Input type to Optical if you are using these.

You also need to make sure that the Clock Source is switched correctly so that Pro Tools synchronizes to the clock signals embedded in the incoming digital audio stream.

note ▷ When you connect any digital audio equipment to any other digital audio equipment, there can only be one clock source or the digital signals will get out of step with each other and you will hear clicks in the audio and a general loss of quality. All digital audio carries clock signals embedded within the digital audio stream so that the other equipment can synchronize to the same clock during transfers between equipment.

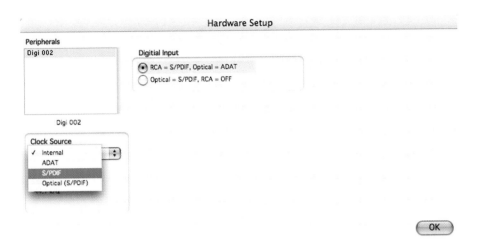

Fig. 4.32 – Switching to external sync.

tip ▷

After you have finished transferring the audio from CD or DAT you should always reset the clock source to Internal. If you don't do this, Pro Tools will continue to synchronize to the external equipment at whatever sample rate this happens to be set to. If this is the same sample rate as your session and subsequent sessions, this may not be a problem. However, if you choose a different sample rate for your next Pro Tools session, you *will* have a problem that you may not notice until some time later when you disconnect or re-configure the external equipment or re-configure your Pro Tools Session.

If the sample rate on your DAT machine reverts to its default of 48 kHz while you are still working with a 44.1 kHz Pro Tools Session, for example, then your Pro Tools Session will be marked as a 44.1 kHz Session but will actually be recorded at 48 kHz. While it remains synchronized to an external clock running at 48 kHz it will sound OK, but when you run it at what Pro Tools 'thinks' is the correct sample rate of 44.1 kHz, the audio will run slow – so the pitch of any music will drop, for instance.

Importing video

Importing video is very straightforward. Just go the File menu and choose the 'Import: QuickTime Movie. . .' command. The 'Open Movie' dialog appears and you can use this to navigate through your disk drives and folders until you find the QuickTime movie that you want to import. Select the movie and click the Open button and the movie will appear in a special movie track in the Edit window. Open the QuickTime Movie from the Window menu to view the movie within Pro Tools LE.

Importing Audio from Video

There are two extremely useful commands in the File:Import sub-menu. The first of these (Audio from Current Movie. . .) lets you import audio from a movie that you have already imported into your current Session while the second (Audio from QuickTime Movie. . .) lets you import audio from any QuickTime movie that you have available to your computer system (on a disk drive, across a network, or whatever).

You are asked to choose a destination folder for the imported audio and you will normally choose the Audio folder in your current Session folder. The imported audio will appear in your Audio Regions list and you can simply drag and drop this onto any suitable empty audio track in your session.

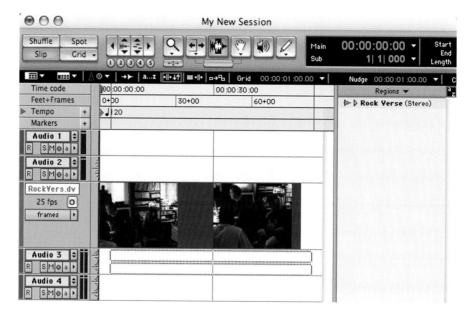

Fig. 4.33 – Placing imported audio from movie.

What this chapter has covered

When you have completed this chapter, you should know how to open a new Pro Tools Session, choose an appropriate sample rate, bit depth and file type, and save the session to a suitable disk drive.

You should also be able to allocate hard disk space for your recordings, label your interface's inputs and outputs, and create audio and MIDI tracks, aux inputs and Master faders.

And if you have existing material to work with, you should now know how to get this into your Session by one method or another.

5

Tempos & Grooves

Pro Tools LE has an excellent set of features for setting and adjusting tempos, matching grooves, and changing the tempos of imported files to match the Session tempo and vice versa.

Setting up Tempo, Meter, and Click

If you intend to do anything but the most basic recording and playback of musical audio material you will need to make sure that the bar and beat positions in the audio line up with the bar and beat positions in Pro Tools LE. When the bars correspond correctly you can navigate to bar positions and use Grid mode and other features to edit your audio.

If you are recording musicians, you can generate a click for the musicians to play along to. If they do this well, then what they play will be in time with the click and will line up with the bar lines in Pro Tools.

Before you set the tempo you will also need to consider whether it will be necessary to change the time signature or meter.

Setting the Meter

Pro Tools LE defaults to 4/4 meter, which is the time signature of most popular music. If you want to work in other meters such as 3/4 (waltz time) or 5/4 (like Dave Brubeck's jazz hit 'Take 5') or with the changing meters that are used in various forms of music such as classical and orchestral film scores, then you will normally need to set these before you start recording.

When the conductor icon is not selected, you can double-click on the displayed time signature in the MIDI controls section of the Transport window to open the Meter Change window.

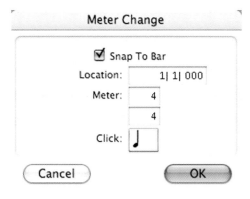

Fig. 5.1 – Meter Change window.

Alternatively, when the Conductor button is highlighted, you can insert meter changes using the Change Meter window that you can select from the Event menu's Time: Operations sub-menu.

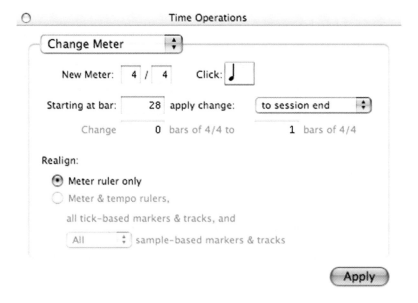

Fig. 5.2 – Time Operations Change Meter window.

Setting Tempos

Manual tempo mode

You can set a tempo for your session manually using the MIDI tempo controls in the expanded Transport window. When the conductor icon is not selected, you can either type the tempo into the tempo field or use the slider to change the tempo.

Fig. 5.3 – Setting the tempo using the MIDI tempo controls.

Tempo events

When you have decided on a tempo for your session, it makes good sense to enable the Conductor track by selecting the Conductor icon in the Transport window and enter the correct tempo for your session into the Tempo Ruler's Song Start Marker.

Step 1. Go to the Tempo Ruler and double-click the red Song Start Marker triangle at the start of the track to open the Tempo Change dialog.

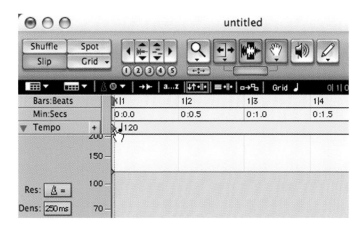

Fig. 5.4 – Clicking on the Song Start Marker in the Tempo Ruler.

Step 2. Type the tempo you want to use for your Session into the BPM field and click OK to replace the default 120 BPM value at the start of the Song.

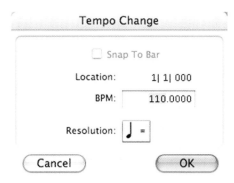

Fig. 5.5 – Tempo Change dialog.

Building tempo maps

You can build a tempo map for the Conductor track by inserting Tempo Events into the Tempo ruler at the locations where you want the tempo to change. You can insert tempo events using the Tempo Operations window that you will find in the Event menu's Tempo sub-menu. There are six Tempo Operations windows to choose from. The first lets you set a constant tempo at a specified bar position or between two specified bar positions. This is the one that you will use most often. To insert Tempo events:

Step 1. Click in the Tempo ruler at the location where you want to insert the tempo event.

Step 2. Click the Add Tempo Change button to the left of the Tempo Ruler to open the Tempo Change dialog.

Step 3. In the Tempo Change dialog you can type the exact location where you wish to add the new tempo (very useful if you just clicked somewhere nearby this to open the dialog) if necessary, then type the new tempo in BPM and click OK.

tip ▷ To save typing the location if you are close to this, you can tick the 'Snap To Bar' option to place the new tempo event exactly on the first beat of the nearest bar.

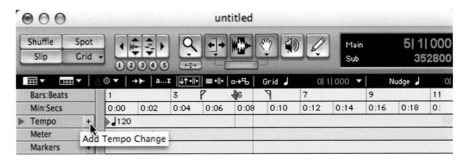

Fig. 5.6 – Adding a Tempo Change to the Tempo Ruler.

Tempo Operations

If you are creating music that requires lots of tempo changes, when working to picture, for example, there are six Tempo Operations windows accessible from the Event menu that you can use. The first of these lets you set a constant tempo over a selected range of measures. The next three windows let you create a linear, parabolic, or S-curved 'ramp' of tempo changes between a pair of specified bar positions. The final two windows let you scale or stretch existing tempos between specified bar positions. Note that each window has a basic set of controls and a more advanced set that is revealed when you check the 'Advanced' check box.

For instance, if you want to change the tempo between Bar 3 and Bar 5, returning to the original tempo after Bar 5, do the following:

Step 1. Select Bar 3 through to Bar 5 by dragging the cursor along the rulers.

Step 2. Open the Constant Tempo Operations window from the Event menu.

Step 3. Type the tempo you want into the Tempo field.

Step 4. Tick the box for 'Preserve tempo after selection'.

Step 5. Click Apply.

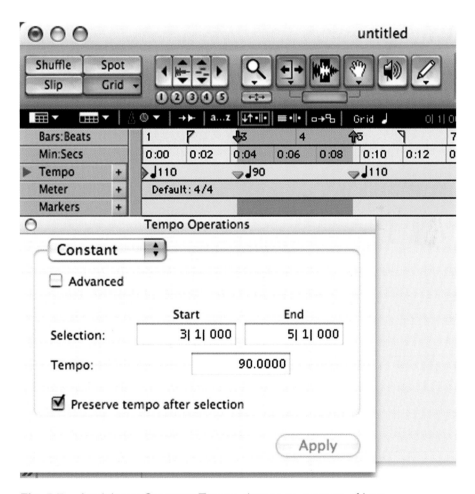

Fig. 5.7 – Applying a Constant Tempo change to a range of bars.

As you can see from the screenshot, the tempo change occurs at Bar 3 and another tempo change back to the original tempo is inserted at Bar 5.

Using the Click

Pro Tools provides a Click feature that lets you use the Digidesign Click plug-in or a virtual instrument plug-in or an external MIDI device to play the click.

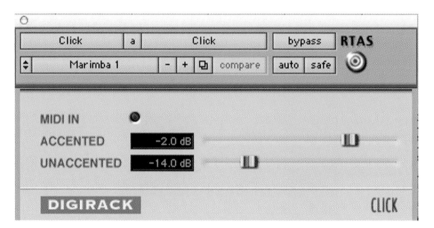

Fig. 5.8 – The Click Plug-in.

To use the Click plug-in you simply need to insert one onto an Auxiliary Input track and make sure that the Metronome button is highlighted in the Transport window or that 'Click' is selected (checked) in the Options menu.

You can use the Click/Countoff Options dialog to choose when the click will be active and to set the MIDI notes, velocities and durations that will play if you are using a virtual instrument or an external MIDI device.

note ▷ The MIDI parameters in the Click/Countoff Options dialog do not affect the Click plug-in.

You can also set the number of count-in bars that you prefer and whether you will hear these every time you playback or only when recording.

note ▷ If you are using the internal Click plug-in, leave the Output popup set to 'None'. If you are using an external device or a software synthesizer to produce your click sound, you can choose this using the Output popup.

Fig. 5.9 – Click/Countoff Options dialog.

If you prefer to use the sound of a drum machine you can either insert a plug-in or connect an external MIDI device to an Auxiliary track and set the output of the Click to play whichever MIDI notes will produce the sound you want. In the example below, I set this up as an alternative to the built-in click sound using a couple of sounds from the classic CR78 beatbox included in the Zero-G Nostalgia plug-in.

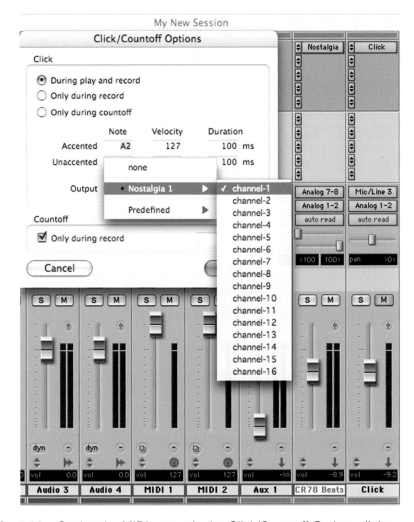

Fig. 5.10 – Setting the MIDI output in the Click/Countoff Options dialog.

note ▷ If you want to record the Click (or any other plug-in that you are using to produce the metronome sound) as audio, you will need to route the output from the Aux track that you are using to monitor the click sound into the input of an Audio track and then record into this.

Adjusting Tempos

If you import existing audio you may need to either adjust the tempo of the imported audio to the Session tempo or adjust the Session tempo to the tempo of the imported audio. Various techniques for adjusting tempos are presented in this chapter.

Adjusting the Session tempo to the tempo of
imported (or recorded) audio

Creating a tempo map using the Identify Beat command is a key area that you
should definitely become familiar with if you want to match the tempo of your
session to the tempo of any audio that you have imported (or recorded without the musicians playing to a click).

To be able to follow any tempo changes and to allow editing to bars and beats,
you have to be able to calculate the tempo of the music at the start and whenever the tempo changes. In other words, you have to create a tempo 'map'
that reveals any tempo changes.

Step 1. First of all, set up a click in Pro Tools and adjust this until it is somewhere near the tempo of the audio you are trying to match to.

Step 2. Make sure that you are in Conductor mode by clicking on the
Conductor icon in the Transport window.

Step 3. Make sure that you have the Bars:Beats and Tempo rulers showing
in the Edit window and open the Tempo Editor so that you can see what is
happening here.

Step 4. Double-click on the small red arrowhead in the Tempo Ruler to bring
up the Tempo Change dialog. Set the tempo to the value you think it should
be and click OK. Then listen to the music with the click. If they are not in step,
keep doing this until they are fairly close in tempo.

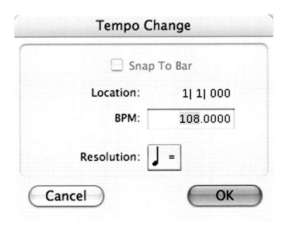

Fig. 5.11 – Tempo Change
dialog.

Step 5. When you have the click roughly matching the tempo of the music,
move the first beat of the music to correspond to Beat 2 in your Pro Tools
Session. This allows for any pick-up notes before the first downbeat and lets
you slip the region back and forth while you are adjusting the positioning. You
can drag the region back and forth with the mouse using the Selector Tool, but
it is a lot easier to use the Plus and Minus keys on the numeric keypad to
nudge the region. I usually set the nudge value to, say, 100 samples until I get

very close. Then I zoom in and set the nudge to 10 samples and eventually down to 1 sample to be really accurate.

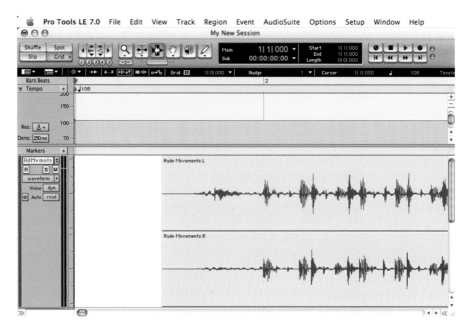

Fig. 5.12 – Edit window showing Bar:Beat ruler, open Tempo Editor, and music lined up with the first downbeat at Beat 2.

To calculate tempos, Pro Tools offers the Identify Beat command. The way this works is that you accurately select a given number of bars, then choose the Identify Beat command from the Edit menu or type Command-I (Control-I in Windows). This brings up a dialog box where you can type in what you have decided the start and end bar numbers are.

note ▶ If you have made your selection of the waveform with sufficient accuracy, and if you have 'told' Pro Tools the correct number of bars and beats, then Pro Tools can easily work out what the tempo in the sequence would have to be for the length of audio that you have selected to correspond to the number of bars and beats that you say it lasts for.

Step 6. So, for example, select 4 bars of music, making sure that it is exactly 4 bars (no more and no less) by playing the selection repeatedly in Loop Playback mode until you are satisfied there are no timing glitches around the loop as you listen to it.

note ▶ You will need to zoom in closely to 'fine-tune' your edit points and make sure they are on zero waveform crossings for the most accurate results.

Fig. 5.13 – The 4 bars selected. Notice that the displayed length says that the selection is 4 bars and 15 ticks in this example. But your ears and your eyes tell you that what you are hearing and seeing is exactly 4 bars of music.

Step 7. When you are satisfied that you have exactly 4 bars selected, use the Identify Beat Command from the Event Menu to bring up the Add Bar | Beat Markers dialog. This will show the Start and End locations according to the tempo currently set. So the Add Bar | Beat Markers dialog will look something like Fig. 5.14.

Step 8. But you know better! You know that you have selected exactly 4 bars of audio that starts at, say, Bar 3 and finishes at the beginning of Bar 7 – your own ears and eyes tell you that. So all you have to do is to make sure that the Locations read 3|1|000 and 7|1|000 by typing these numbers into the location fields and hit the OK button – see Fig. 5.15.

Fig. 5.14 – The Add Bar | Beat Markers dialog.

Fig. 5.15 – Defining the selected number of bars in the Bar | Beat Markers dialog.

Step 9. Pro Tools then works out what tempo changes are needed to make the session show your selection as 4 bars exactly between bars 3 and 7 and inserts these into the tempo map – as in Fig. 5.16.

Fig. 5.16 – The 4-bar selection now matches the bar lines correctly.

Step 10. To create a tempo map for the entire piece of music, you have to repeat this procedure for each section in which the tempo changes – bar by bar if necessary.

note ▶ If the music that you are tempo mapping has lots of tempo variations, it could easily take you two or three hours or more of intensive work to create an accurate tempo map.

If you are following this example on your computer, your Pro Tools session should now look similar to Fig. 5.17.

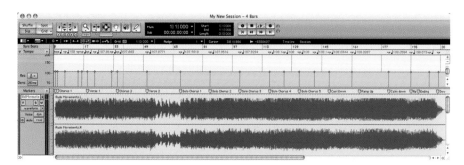

Fig. 5.17 – A tempo mapped Pro Tools Session.

Adjusting the tempo of imported (or recorded) audio to the session tempo

If you are working on dance music, a typical way to get started is to import some ready-made drum beats that you can build stuff around.

For example, I recently acquired a library of full of Apple Loops for GarageBand called Dance Pack 2. I looked in a folder called 'Beats "N" The Hood' and found a 4-bar audio file called 'Thursday-Dry 1'. This was marked as 107 BPM, which was close enough to the 108 BPM of the Session I was working on. I opened the Workspace Browser, dragged this from the DVD-R and dropped it onto a track in Pro Tools.

Fig. 5.18 – Drag 'n' drop audio from the Workspace Browser onto a track.

Four bars of audio running at 107 BPM lasts just a little longer in time than 4 bars running at 108 BPM. So I needed to adjust the tempo of this imported audio to the session tempo.

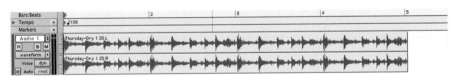

Fig. 5.19 – Look closely to the right of the screenshot at where the region ends in relation to the bar marking in the ruler above and you will see that this 4-bar audio file running at 107 BPM lasts just a little longer than 4 bars running at 108 BPM would. (4 bars running at 108 BPM would end exactly at the ruler mark for Bar 4.)

Here's the trick:

Step 1. Using Grid Mode with the grid set to 1-bar resolution, switch the Main Counter to Bars:Beats and using the Selector Tool select exactly 4 bars on the timeline with the tempo set to 108 BPM – leaving the small amount of the audio region in Bar 5 unselected.

Step 2. Change the Main Counter to show Samples and note how many samples 4 bars of audio lasts at 108 BPM. You might find it convenient to view Comments in the Edit window and write the value there.

Fig. 5.20 – Note how many samples for 4 bars at 108 BPM.

Step 3. Switch to the Grabber Tool and select the whole audio region.

Fig. 5.21 – Select the whole audio region.

Step 4. Go to the AudioSuite menu, open the Time Compression Expansion plug-in, type the number of samples occupied by 4 bars at 108 BPM into the Destination field and press Return on your computer keyboard to confirm this.

Step 5. Set the Accuracy slider to Rhythm (hard right) for best results with drums or percussion.

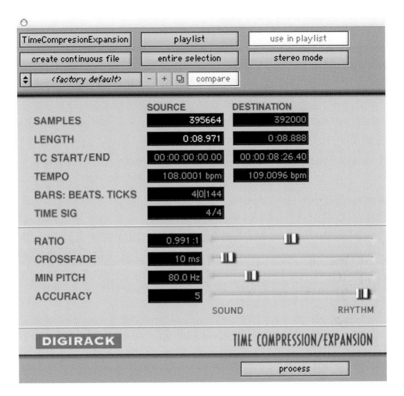

Fig. 5.22 – Type the number of samples in the Destination field and set the Accuracy slider.

Step 6. Make sure that 'Use in Playlist' is selected at the top right of the plug-in. This ensures that a new processed file is created that replaces the original file in the playlist in the track. The original file is left untouched and is still available in the Audio Regions list, so you can always re-do the processing if anything goes wrong.

Step 7. Finally, click on the Process button to create a Time Compressed file that exactly fits 4 bars at 108 BPM.

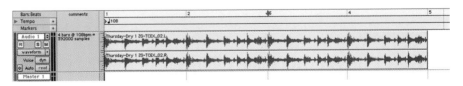

Fig. 5.23 – Exactly 4 bars at 108 BPM.

Now you can easily create a section for your new song by selecting the 4 bars and using the Repeat command from the Edit menu, entering '7' to add a further 28 bars to make up a 32 bar section, for example.

Fig. 5.24 – Repeat 7 times to form a 32-bar section.

What is Beat Detective?

Beat Detective is a set of useful tools that you can use to get everything lined up correctly in your Session. As the documentation explains, Beat Detective analyzes and corrects timing in performances that have strong transient points, such as drums, bass, and rhythm guitar. It allows the user to define a tempo map from a performance or to conform the performance to a tempo map by separating it into regions and aligning it to the beats. But isn't that what we just did using the Identify Beat command? Well, the manual describes it in a kind of fancy way, but it does mean that you can use Beat Detective to make the Session tempo match the tempo of imported or recorded audio, or to make the tempo of imported or recorded audio match the session tempo. The difference is that Beat Detective promises to save you hours of endless fiddling around with the Identify Beat command – and sometimes it can.

So why not use Beat Detective?

You may be wondering whether to use Beat Detective to work out the tempos instead of using the Identify Beat command, especially if there are a lot of tempo changes. After all – that's what Beat Detective is supposed to do, right? Well, maybe... Beat Detective can work out a tempo map automatically for you, but it works best with fairly straightforward and well-defined rhythmic sections. With short sampled drum loops with very clearly defined beats, Beat Detective works great. Eight or sixteen bar sections of a fairly steady 'live' recording can also work well. But if you have a long piece of music and especially if the beats are not so clearly defined, then your ear will make a much better job of analyzing where the beats are than Beat Detective could ever do. Also, if there is only a handful of tempo changes throughout the music, it can be quick enough to use the Identify Beat command – sometimes even quicker than using the Beat Detective.

note ▷   The version of Beat Detective available in Pro Tools LE does not have the Collection Mode that is used when editing multi-mic'ed drum kits in the TDM version. The optional Music Production Toolkit adds this feature.

OK, let's use Beat Detective!

Beat Detective identifies the individual beats in your audio selection by looking for the peaks in the waveform. You can adjust the settings until you have identified most of these and then edit manually to fine-tune the choices. The points identified are referred to as 'beat triggers' in Pro Tools and these can be converted to Bar|Beat markers. Once the beats have been identified correctly, Beat Detective can extract the tempo from the audio – creating a tempo map. Other audio regions and MIDI tracks can then be quantized to these markers.

So, for example, you can use Beat Detective to extract tempo from audio that was recorded without listening to a click – even if the audio contains varying tempos, or material that is swung – and you can then quantize other audio regions or MIDI tracks to this 'groove'.

You can also do the opposite of this: if your session already has the right tempo, you can use Beat Detective to 'conform' any audio with a different tempo (or with varying tempos) to the Session's tempo. You can choose to keep a percentage of the original feel if you like, and you can increase or decrease the amount of swing in the conformed material. You can also conform regions which you have previously separated using Beat Detective to a Session's tempo map.

First you define a selection of audio material on a single mono or multi-channel track, or across multiple tracks. Then adjust the Detection parameters so that vertical beat triggers appear in the Edit window, based on the peak transients detected in the selection. For example, with a 'boom-chick' bass drum and snare drum beat you would see a vertical line in the display immediately before each bass drum and snare drum beat. You should examine these triggers visually to make sure that there are none in the wrong places for any reason. When you are satisfied that the triggers look OK, you can generate Bar|Beat Markers based on these beat triggers to form a tempo map from the selection which you can use for your session.

One of the most useful applications is aligning loops with different tempos or feels. If one loop has a subtly different feel or groove you can use Beat Detective to impose that groove onto another loop. This is great for remixes where you often need to extract tempo from the original drum tracks, or even from the original stereo mix. New audio or MIDI tracks can then be matched timing-wise to the original material, or the original material can be matched to the new tracks.

Beat Detective Features

Beat Detective lets you separate and automatically create new regions, representing beats or sub-beats, based on the beat triggers. You can then conform these new regions to the session's existing tempo map or to a groove template.

The Beat Detective window is divided into three sections. The first lets you select the Mode, the next lets you define and capture the Selection, while the third section changes according to the Mode selected. It contains the Detection parameters in the first three modes and swaps these for the Conform

123

parameters and the Smoothing parameters in the last two modes. There are five different modes altogether, as explained below:

Bar | Beat Marker Generation automatically generates Bar | Beat Markers corresponding to transients detected in the audio selection.

Groove Template Extraction extracts rhythmic and dynamic information from the audio and puts this information onto the Groove Clipboard or lets you save it as a DigiGroove template.

Region Separation automatically separates and creates new regions based on transients detected in the audio selection.

Region Conform conforms all separated regions within the selection to the current tempo map. You can preserve some of the original feel of the material with the Strength and Exclude Within option, or impose an amount of swing with the Swing option. Beat Detective can also conform audio regions to groove templates – including its own DigiGroove templates. Cubase, Feel Injector, Logic and MPC style groove templates are provided as standard and you can add your own grooves to the list as you create these.

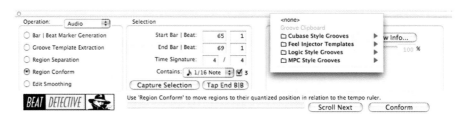

Fig. 5.25 – Beat Detective Conform to Groove.

After conforming regions, gaps may be left between these, so an Edit Smoothing feature is provided which can fill the gaps between regions – automatically trimming them and inserting cross-fades as required. This can save you an awful lot of detailed editing work that would otherwise be necessary to avoid pops and clicks at the region boundaries that would normally need to be trimmed and cross-faded. It also has the advantage of preserving the ambience throughout the track to keep this constant despite the edits.

Using Beat Detective to adjust the Session tempo to the tempo of an imported audio file

As an alternative to the Identify Beat command, Beat Detective can be used to calculate the tempo of an audio region.

Let's see how it works when you want to adjust the Session tempo to the tempo of an imported audio file. I chose the same 107 BPM loop from Dance Pack 2 called 'Thursday Dry-1' and placed this into my Pro Tools Session which was running at 108 BPM. I already knew that this loop-able audio file (Thursday Dry-1) was in 4/4 meter and that it was exactly 4 bars long.

You may not already know the meter and number of bars in the audio file that you choose to work with. In this case, count the number of bars as you audition the audio in Pro Tools. Then check in case the meter is anything other than 4/4 and use the Meter Change dialog to change the meter as necessary.

note ▷

If you don't know the meter of your audio region, then you must listen to it and work this out. If you are a musician you will usually find this easy enough. If not, you will either have to learn the basics of music or ask a musician to help you. You must also make sure that you have selected an exact number of bars – zooming into sample level or using the Tab to Transient feature to check that you make your region selection immediately before the first beat and after the last beat. Again, ask a musician if you are not sure how many bars you have selected.

To adjust the Session tempo to the tempo of an imported audio file:

Step 1. Place your imported audio file at the start of an audio track in Pro Tools and select this in the Edit window. Play the audio repeatedly using Loop Playback mode to check that you have selected an exact number of bars.

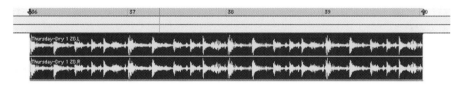

Fig. 5.26 – A 4-bar region of audio running at 107 BPM placed in a Session running at 108 BPM last slightly longer than 4 bars in the Session – look closely at the right of the selection to see this.

Step 2. Open Beat Detective and select Bar | Beat Marker Generation.

Step 3. Click Analyze.

Step 4. Make sure the Sensitivity is set to zero.

Fig. 5.27 – Beat Detective Bar | Beat Marker Generation.

Step 5. Click Generate.

Fig. 5.28 – Realign Session dialog.

Step 6. Decide whether to Preserve the Tick Position or the Sample Position in the Realign Session dialog that appears and click OK.

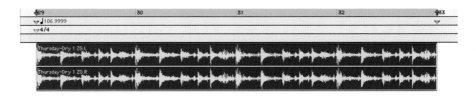

Fig. 5.29 – Tempo is calculated and Bar|Beat Markers inserted.

That's it! The Tempo is calculated and Bar|Beat Markers are placed at the beginning and end of the selection to indicate the new tempo and the meter.

note ▷ In this example, the region was placed to start at the beginning of Bar 29 before using Beat Detective and Beat Detective was instructed that the region should start at the beginning of Bar 29 – so the region would not move anyway. As the Session was being used for music production using Bars:Beats:Ticks to identify locations, the correct option to choose in the Realign Session dialog would be Preserve Tick Position. If the region had actually started before or after Bar 29 (because it had not been precisely positioned in the Edit window), then it would have been moved to the beginning of Bar 29.

Using Beat Detective to adjust the tempo of an imported audio file to the Session tempo

Instead of using time compression or expansion to adjust the tempo of an imported audio file to the Session tempo, you can use Beat Detective to slice the audio into regions containing beats then force these to align with the beats in the tempo map.

Again, I chose the same 107 BPM 4-bar audio file in 4/4 meter from Dance Pack 2 called 'Thursday-Dry 1' and placed this into my Pro Tools Session, which was running at 108 BPM.

You can try this using any audio file containing 4 bars of music. Here's how to change an imported audio file to match the tempo map of the session:

Step 1. Select the audio file in the Edit window, zooming in to make sure that you start exactly at the beginning of the first beat and finish at the end of the last beat choosing your edit points on zero waveform crossings.

note ▷ Your selected region should not encompass any Pro Tools meter or tempo changes. If it does, Beat Detective will not work correctly.

Step 2. Open Beat Detective and choose Region Separation mode.

Step 3. If the tempo and meter of the selection don't match those of your Session, you need to enter the time signature and the start and end Bar:Beat locations. For example, if you have a 4-bar selection starting at Bar 36, type 36 as the Start location and 40 as the End location. This encompasses Bars 36, 37, 38, and 39 and stops at the beginning of Bar 40 (so Bar 40 is not included in the selection). If the time signature is not 4/4, you need to enter this also.

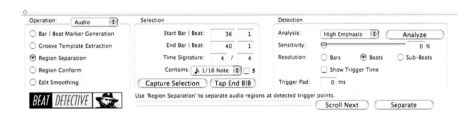

Fig. 5.30 – Beat Detective region separation.

Step 4. Click Analyze.

Step 5. Set the Resolution to Bars, Beats or Sub-Beats according to the complexity of the rhythmic pattern in the audio.

Step 6. Adjust the Sensitivity until beat triggers appear in the selected audio.

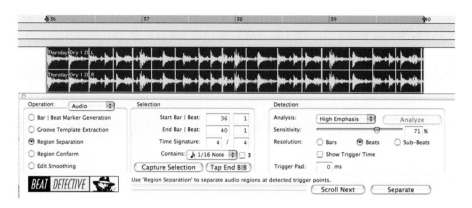

Fig. 5.31 – Generating Beat Triggers.

Step 7. Check that the triggers are falling in sensible places by zooming the waveform so that you can see the individual triggers clearly, using the Scroll Next button in Beat Detective to jump to each trigger in turn.

tip ▷ Don't forget that you can always press Shift-Spacebar or Shift-click the Play button to play back at half speed when you want to check how accurate the beat markers are. This way, the audio plays back slowly enough for you to get a better idea as to whether the Beat Triggers are positioned suitably or not.

Step 8. Make any minor adjustments to the positioning of the Beat Triggers using the Grabber Tool to drag them to the left or to the right.

Step 9. Delete any false triggers by Alt-clicking these.

Step 10. Insert any missing triggers by clicking where you want to insert a missing trigger using the Grabber Tool.

Step 11. Click the Separate button to separate the beats into individual regions based on the detected Beat Triggers.

Fig. 5.32 – Separated Regions.

After you have separated the audio into regions, you can either conform these to the session's current tempo map or to a Groove Map. Here we will use Standard Conform.

Step 12. Select Region Conform mode in Beat Detective.

Step 13. Adjust the Strength, Exclude Within and Swing parameters then click Conform. The regions will now align with the tempo map.

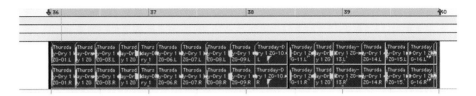

Fig. 5.33 – Regions conformed to Session tempo.

Step 14. Switch to Edit Smoothing mode then click Smooth to fill in any gaps between the regions to prevent clicks.

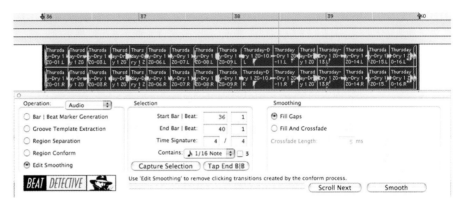

Fig. 5.34 – Gaps between regions filled in by smoothing.

note ▷ Beat Detective works best with rhythmic tracks such as drums, bass, and guitars – so don't expect great results with more ambient material, strings, vocals, or suchlike. And if you have a rhythm track that is too complex or too far out of time, you won't get good results either.

Using REX and ACID Files

Perhaps the best news for anyone who likes to work with audio loops is that Pro Tools 7 imports and plays REX and ACID files.

REX files are created using Propellerhead's ReCycle software. They contain audio 'loops' edited into small slices and stored in a type of AIFF file along with information about the original loop tempo. You can play these back using software such as Reason or Cubase SX and if you change the tempo, these audio 'loops' immediately follow the tempo changes without any need for time-stretching or other processing. Pro Tools audio tracks can do this if you change the timebase from sample- to tick-based. Simply drag and drop a REX file from a DigiBase browser or from the Macintosh Finder to a track, the Track List, the Timeline, or the Region List. The REX file is converted to a region group on import, and all its slices are converted into individual audio files and regions.

ACID files are a 'stretchable' and 'tunable' format created using Sony's ACID software for the PC. They contain metadata about tempo, number of bars, pitch, and slice information along with the audio in a type of WAVE file. Like REX files, you can simply drag and drop ACID files from a DigiBase browser or from the Desktop to the Timeline, a track, the Track List, or the Region List. And you need to change the timebase from sample- to tick-based. Unlike REX files, ACID files don't always contain slice metadata, so you may need to manually slice the imported file using the Separate Regions At Transients command or Beat Detective. If slice metadata is present in the ACID file, it is converted and imported as a region group, and all slices are converted into individual audio files and regions. If slice data is not available, it is imported as a regular audio file.

Typically, REX and ACID files will be exactly 1-, 2-, 4-, or 8-bars long and are used as basic 'building blocks' for musical arrangements. To build up 16- or 32-bar verses, choruses or other sections of the musical arrangement, they are repeated or 'looped'. Often they contain drumbeats, but can also contain basslines, keyboard pads, guitar licks, or any type of sounds.

To Import a REX or ACID File

You can drag and drop a REX or ACID file into an existing audio track, but you must change the Time Base Selector to Ticks or the audio will stay at the original tempo when you change the tempo in Pro Tools.

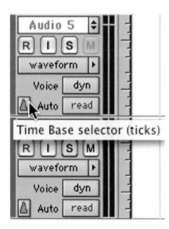

Fig. 5.35 – Click and hold the Time Base Selector to open the popup selector.

Step 1. Put an Audio track into 'ticks' mode.

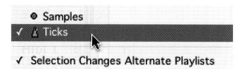

Fig. 5.36 – Choose 'Ticks' from the popup Time Base Selector.

Step 2. Drag and drop the REX or ACID file into the track in the Edit window.

Now you can make the tempo faster or slower and the audio in the REX or ACID file will follow the tempo changes.

note ▶ If you drag and drop a REX or ACID file into the empty space in a new session or below any existing tracks in the Edit window, Pro Tools automatically creates a new track for this and sets it to tick-based.

What this chapter has explained

This chapter is extremely important for anyone who wants to bring existing audio into a Pro Tools Session and have the bar lines in Pro Tools correspond to those in the audio. It is equally important for anyone who records audio without a click and then wants to be able to edit the audio by referring to the bars and beats.

How to adjust the Session tempo or the tempo of existing audio is fully outlined. Beat Detective is introduced and discussed, and detailed instructions are given on how to use this. Finally, ACID and REX files are explained and instructions are given about using these in Pro Tools.

As with most software features, you need lots of practice if you want to become sure-footed, speedy and confident with these techniques.

6

Recording MIDI and Using Virtual Instruments

A popular way to start a project, whether a song or an instrumental, is to set up a drum pattern, a bassline and a keyboard 'pad' to mark out the chord changes – often using MIDI instruments, although you can always put guide parts down from real instruments as audio.

This is particularly important if you intend to overdub live musicians one at a time. Most musicians prefer to hear something more interesting than a click to play to while overdubbing, and having a keyboard pad throughout helps to guide the musicians through the arrangement. Drums and bass are arguably even more important than lyric, melody and harmony for many types of dance music, so most dance music starts out with these, and even if you will be recording mostly (or even all) live instruments, it often helps if you sketch the arrangement out first using MIDI instruments.

MIDI Setup

If you are going to use MIDI with Pro Tools, you need to set up your MIDI connections first.

The MIDI sub-menu in Pro Tools LE's Setup menu contains four items: MIDI Studio, MIDI Beat Clock, Input Filter, and Input Devices.

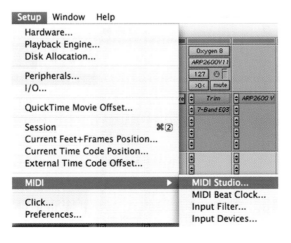

Fig. 6.1 – Setup Menu showing the MIDI sub-menu.

MIDI Studio. . .

With Mac OS X, you can use Apple's Audio MIDI Setup (AMS) utility to iden-tify which external MIDI devices are connected to your MIDI interface and to configure your MIDI studio for use with Pro Tools.

Step 1. Open the AMS utility by choosing MIDI Studio. . . from the Setup menu's MIDI sub-menu.

There are two pages, one for Audio Devices and one for MIDI devices – each with its own selection tab at the top of the window.

Step 2. Click the MIDI Devices tab.

The AMS utility scans your system for connected MIDI interfaces. When it finds your interface, this appears in the window with each of its ports num-bered. Then you need to add devices corresponding to the actual devices that are connected to your MIDI interface.

Step 3. Click Add Device.

A new external device icon with the default MIDI keyboard image appears. You can drag this icon around the screen to place it wherever you find most convenient.

Step 4. Connect the MIDI device to the MIDI interface by clicking the arrow for the appropriate output port of the device and dragging a connection or 'cable' to the input arrow of the corresponding port of the MIDI interface.

In this example, the Oxygen 8 keyboard controller only transmits MIDI data to the Digi 002, it doesn't receive any data itself. So all you need to do is to drag a 'cable' from the outward pointing arrow on the Oxygen 8 icon to the inward pointing arrow on the Digi 002 icon to represent the connection.

tip ▷ Don't forget that you need to connect a physical MIDI cable from the Oxygen 8 to the rear panel of the Digi 002 in this scenario. (It is also possible to con-nect the Oxygen 8 directly to the Mac via USB so that it acts as its own MIDI interface.)

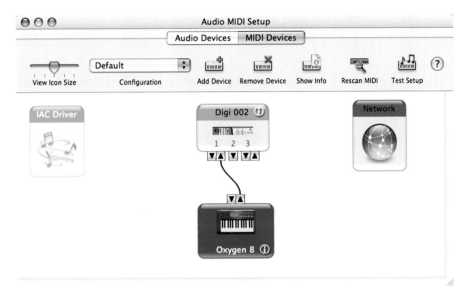

Fig. 6.2 – MIDI Studio – A typical Setup using an Oxygen 8 keyboard with a Digi 002.

Step 5. Double-click on the icon of any device that you have inserted (such as the Oxygen 8 in this example).

A dialog window opens to let you set various parameters for the device. If the device is a popular model that has been available for some time, it will be listed in the Model popup menu in this dialog.

Step 6. Select the Manufacturer and Model of the connected device so that the parameters are automatically set correctly for you. Alternatively, if the device is not listed, just type whatever names you like for these and set the Transmits and Receives and other parameters manually.

Step 7. Using the Icon Browser, you can choose from a selection of icons to represent the MIDI device – such as a generic MIDI keyboard.

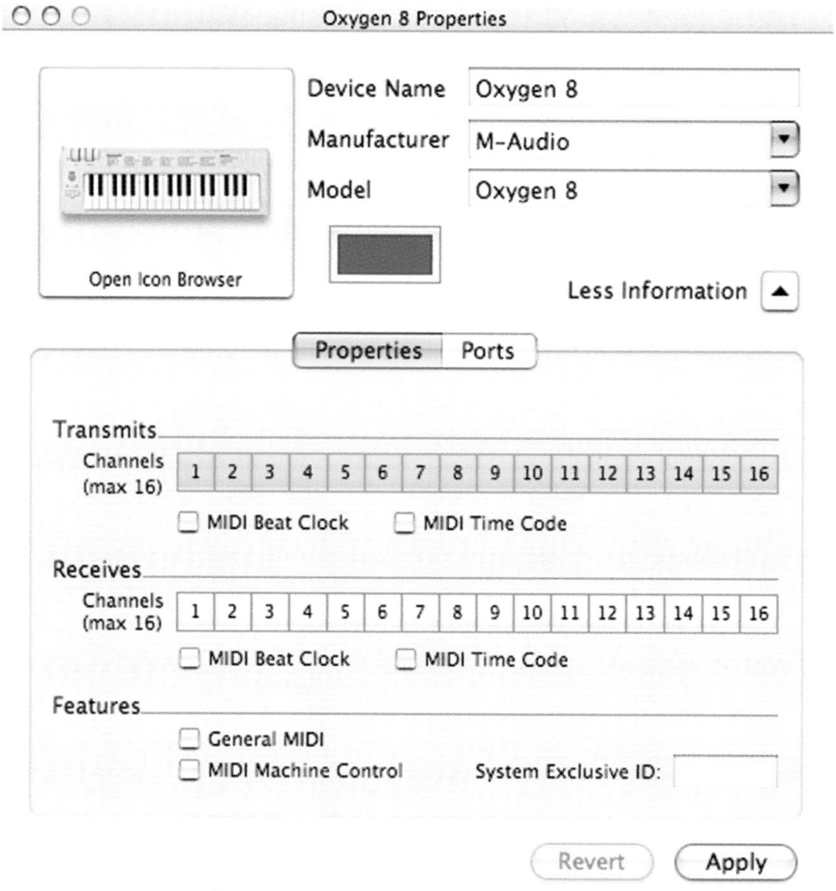

Fig. 6.3 – MIDI Studio Properties dialog with Device Properties Tab revealed.

Step 8. Click the Apply button to apply your settings.

MIDI Beat Clock. . .

You may have MIDI devices connected to your Pro Tools system that can use MIDI Beat Clock to synchronize with the tempo of the Pro Tools sequence. This is often the case with drum machines or with synthesizers that have arpeggiators. Delay units are another example where the timing of the delay repeats can be synchronized to the sequence tempo.

Many plug-in instruments and effects can be synchronized. Some of these plug-ins automatically configure to receive MIDI Beat Clock from Pro Tools. Plug-ins that do not self-configure are detected when inserted, then listed in the MIDI Beat Clock dialog where you can select the ones that you wish to synchronize manually.

As you will see from the screenshot, all the Reason modules are listed here because I inserted Reason before opening this dialog.

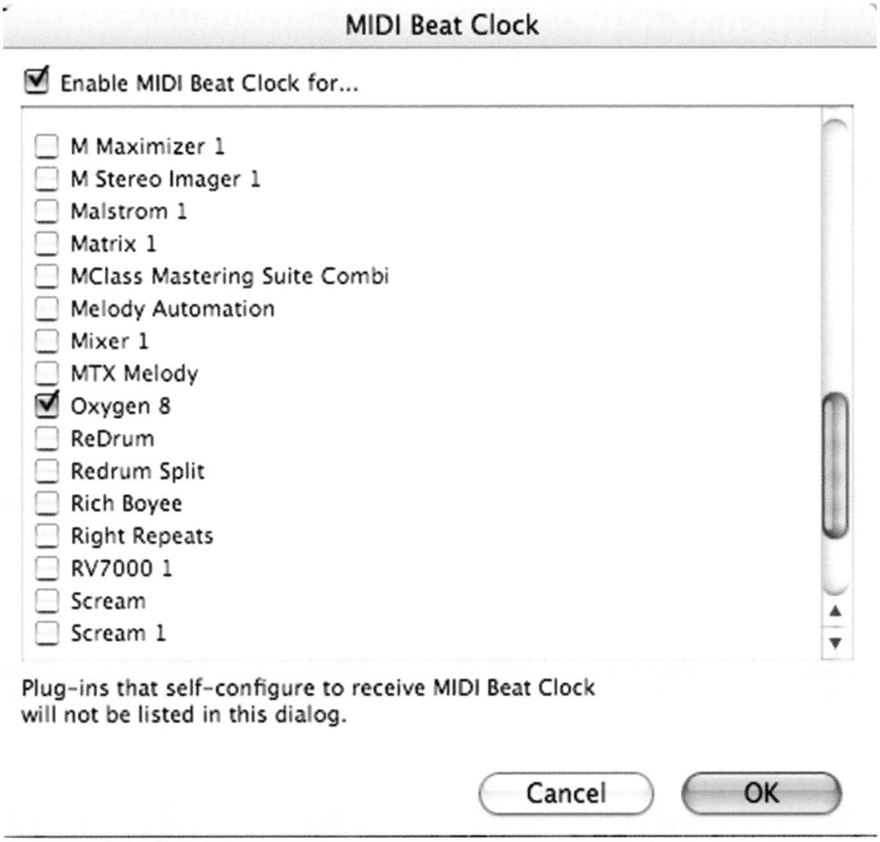

MIDI Beat Clock

☑ Enable MIDI Beat Clock for...

- ☐ M Maximizer 1
- ☐ M Stereo Imager 1
- ☐ Malstrom 1
- ☐ Matrix 1
- ☐ MClass Mastering Suite Combi
- ☐ Melody Automation
- ☐ Mixer 1
- ☐ MTX Melody
- ☑ Oxygen 8
- ☐ ReDrum
- ☐ Redrum Split
- ☐ Rich Boyee
- ☐ Right Repeats
- ☐ RV7000 1
- ☐ Scream
- ☐ Scream 1

Plug-ins that self-configure to receive MIDI Beat Clock will not be listed in this dialog.

(Cancel)   (OK)

Fig. 6.4 – MIDI Beat Clock dialog.

MIDI Input Filter. . .

The MIDI Input Filter dialog lets you filter out any MIDI data that you don't want to record.

For example, very few MIDI devices can respond to Polyphonic Aftertouch data, but you may have a keyboard such as the Yamaha DX1 that produces this data when you play its keyboard (I used to own one of these myself). There is no point in recording this kind of data unless you will be using it to play sounds (such as a few of those in the DX1) that can respond to it. So you should filter this out.

For similar reasons, and to help while troubleshooting your MIDI system, you may need to filter out other types of MIDI data, so Pro Tools provides lots of filter choices in its MIDI Input Filter dialog.

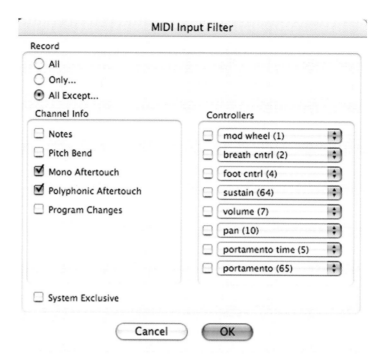

Fig. 6.5 – MIDI Input Filter dialog.

MIDI Input Devices. . .

If you have several MIDI devices attached to your Pro Tools system, it can be useful to disable the MIDI input from any that you are not using to avoid recording unwanted MIDI data that may be 'flying around' your MIDI system.

Just imagine that you have two keyboards connected and, while you are recording using one, your cat jumps on the keys of the other! If you disable the input from this unused keyboard, you won't get any unexpected surprises.

Fig. 6.6 – MIDI Input Devices Enable dialog.

Getting Started

You should check the settings in the MIDI section of the Preferences window before you start recording. For example, some synthesizers and samplers set Middle C to correspond to the C designated as C3 while others set this to C4. Pro Tools allows you to choose whichever.

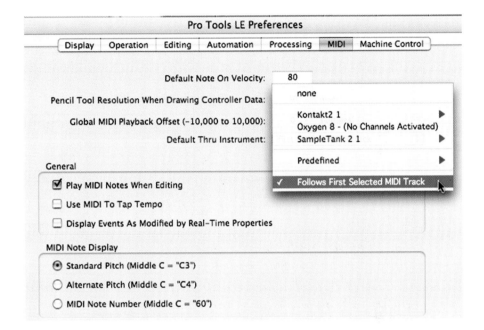

Fig. 6.7 – MIDI Preferences.

If you set the Default Thru Instrument to 'Follows First Selected MIDI Track', then all you need to do to play any MIDI instrument is to click on the MIDI or Instrument track name in the Mix or Edit window and play a connected MIDI keyboard or other device. MIDI Thru must be selected in the Options menu for this to work, and your MIDI device or virtual instrument must be set up so that it is successfully sending MIDI data to Pro Tools.

Take a look at Fig. 6.8 to see how the Mix window looks with a connected Oxygen 8 keyboard playing a SampleTank 2 virtual instrument inserted onto a Pro Tools LE Instrument track. As soon as you select (click to highlight) the Track Name, you can play the Oxygen 8 and it plays Sample Tank 2.

If the previous instrument, Instrument 1 in the screenshot, were selected, this would play instead of SampleTank, because Instrument 1 would then be the first selected MIDI track (i.e. the leftmost in the Mix window).

Fig. 6.8 – Default Thru set to play the first selected MIDI track – in this case, SampleTank.

Setting up to Record MIDI onto a track

Step 1.   Choose 'Input Devices' from the MIDI sub-menu in the Setup menu and make sure that your input device is selected in the MIDI Input Enable window.

Step 2.   Choose 'New. . .' from the Track menu and specify 1 MIDI Track, then click Create.

Step 3.   In the Mix window, click the track's MIDI Input Device/Channel Selector and assign a device and channel from the popup menu that you will use to play MIDI data into Pro Tools. If you have more than one device hooked up that you want to use to create MIDI input, an option is provided to enable 'All' connected MIDI devices for input.

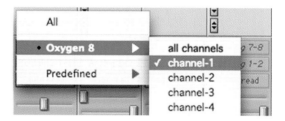

Fig. 6.9 – Using the MIDI Input Channel Selector.

Step 4.   You can assign a default program change to the track by clicking the Program button in the Mix window to open the Patch Names window.

Fig. 6.10 – Clicking the Program button on a MIDI track in the Mix Window.

In the Patch Names window, make the necessary selections for program and bank select, and then click 'Done'. The selected program and bank change messages will be sent when the track plays back.

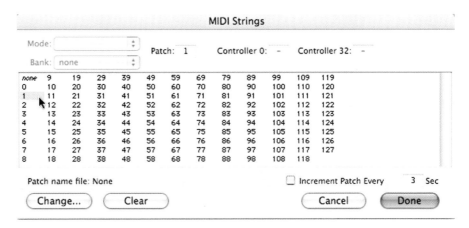

Fig. 6.11 – Patch Names window.

Step 5. In the Mix window, record-enable the MIDI track.

Step 6. Make sure that MIDI Thru is selected in the Options menu.

Fig. 6.12 – Record Enabling a MIDI track.

Fig. 6.13 – Selecting MIDI Thru in the Options Menu.

Step 7. Now when you play some notes on your MIDI controller, you should hear the sound of the MIDI instrument assigned to the track and you should see the track's meter indicating MIDI activity.

Fig. 6.14 – Track Meter showing MIDI activity.

note ▷ The above steps all apply when recording MIDI onto Instrument tracks. The main difference is that the MIDI Input and Output popup selectors are located at the top of the Instruments section rather than in the usual track I/O section.

Recording onto a MIDI or Instrument track

Step 1. Record-enable the track: make sure that the MIDI or Instrument track that you want to record onto is record-enabled and receiving MIDI. The track meters will indicate MIDI input when you play your MIDI device if everything is set up and working OK.

Step 2. In the Transport window, click Return To Zero to start recording from the beginning of the session – or simply hit the Return key on the computer's keyboard.

tip ▷

> Alternatively, you can start recording from wherever the cursor is located in the Edit window. Or, if you select a range of time in the Edit window, recording will start at the beginning of this selection and will automatically finish at the end of the selection.

Step 3. Click Record in the Transport window to enable record mode.

Step 4. Click Play in the Transport window or press the Spacebar to actually begin recording.

tip ▷

> You can start recording by pressing and holding the Command key (Control key on the PC) then pressing the Spacebar on your computer keyboard – which is even faster than using Steps 3 and 4 above. And if you prefer to hit a single key, just press function key F12, or press 3 on the numeric keypad (when the Numeric Keypad Mode is set to Transport) instead.

Fig. 6.15 – Transport Window with Record and Play buttons engaged.

note ▷

> Remember to select the Wait for Note icon if you want the first MIDI note that you play to start the recording. If you want to hear a metronome click while you record, make sure that the click is set up correctly and that the metronome icon is selected in the Transport window. And if you want a countoff, make sure that the Countoff icon is selected.
>
> If you are using Wait for Note, the Play, Record, and Wait for Note buttons flash and recording will start when the first MIDI event is received.
>
> If you are using Countoff, when you click Play, the Record and Play buttons will flash during the Countoff, then recording will begin.

Step 5. Play your MIDI instrument.

Step 6. When you have finished recording, click Stop in the Transport window, or press the Spacebar. The newly recorded MIDI data will appear as a MIDI region on the track in the Edit window and in the Regions List.

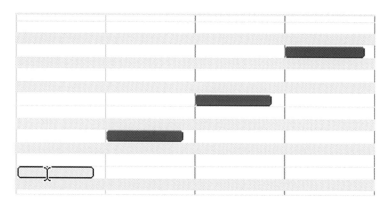

Fig. 6.16 – MIDI Data in the Edit Window.

Loop recording MIDI

There are two ways to loop record with MIDI – either using the normal non-destructive Record mode with Loop Playback and MIDI Merge enabled for drum-machine style recording, or using the special Loop Record mode to record multiple takes on each record pass – as when loop recording audio.

To set up drum-machine style loop recording, where each time around the loop you record extra beats until you have constructed the pattern you want, you need to enable the MIDI Merge function by clicking on the icon at the far right in the Transport. This looks a bit like a letter 'Y' on its side. Deselect 'QuickPunch', 'Loop Record', and 'Destructive Record' in the Operations menu, but select 'Loop Playback' so that the loop symbol appears around the 'Play' button in the Transport window. Select 'Link Edit and Timeline Selection' from the Operations menu, then make a selection in the Edit window to encompass the range that you want to loop around. If you want to hear the audio that plays immediately before the loop range as a cue, you will need to set a Pre-Roll time. So, for example, you would hear your session play back from, say, 2 bars before the loop range, then it would play around the loop until you hit 'Stop'. Each time through the loop you can add more notes until it sounds the way you want it to, without erasing any of the notes from previous passes through the loop – just like with a typical drum machine.

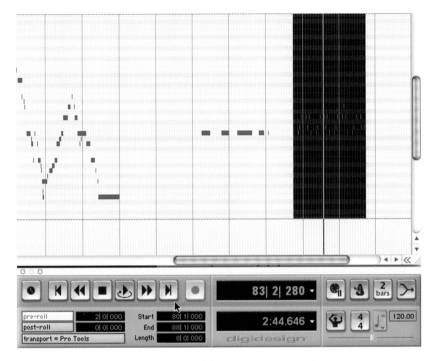

Fig. 6.17 – Loop Recording: Drum-Machine style with Loop Playback and MIDI Merge enabled.

If you record MIDI using the Loop Record mode instead, new regions are created each time you record new notes during successive passes through the loop.

This time, select 'Loop Record' and 'Link Edit and Timeline Selection' from the Operations menu, deselect 'Loop Playback' and check that 'QuickPunch' and 'Destructive Record' are disabled. With Loop Record enabled, a loop symbol appears around the 'Record' button in the Transport window. Make a selection in the Edit window to encompass the range that you want to loop around and set a Pre-Roll time if you need this. Start recording and play your MIDI keyboard or other MIDI controller. Each time around the loop, a new MIDI region is recorded and placed into the Edit window, replacing the previous region. When you stop recording, the most recently recorded of these 'takes' is left in the track, and all the takes appear as consecutively numbered regions in the MIDI Regions list.

The easiest way to audition the various takes is to make sure that the 'take' currently residing in the track is selected, then Command-click (Control-click in Windows) on the selected region with the Selector tool enabled. A popup menu appears called the 'Takes List'. This contains all your recorded 'takes'. Choose whichever you like to replace the 'take' that currently appears in the track.

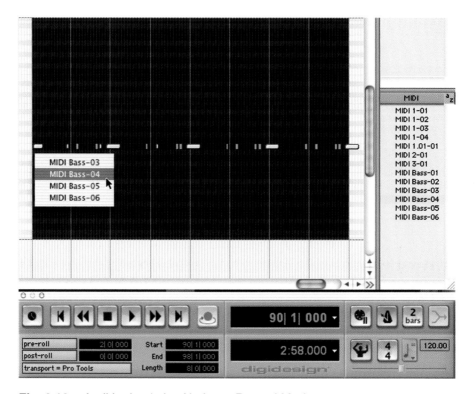

Fig. 6.18 – Auditioning 'takes' in Loop Record Mode.

Playing back a MIDI or Instrument track

Step 1. If you have just finished recording onto the track, click the Record Enable button to take the track out of Record mode.

Step 2. In the Transport window, click Return To Zero to play back from the beginning of the track – or simply hit the Return key on the computer's keyboard.

tip ▷

> Alternatively, you can play back from wherever the cursor is located in the Edit window. Or, if you select a range of time in the Edit window, playback will start at the beginning of this selection and will automatically finish at the end of the selection.

Step 3. Click Play in the Transport window to begin playback. The recorded MIDI data will play back through the track's assigned instrument and MIDI channel and you should hear sound from the connected MIDI device. The popup MIDI Output selector underneath the MIDI Input selector lets you choose which of the available instrument and MIDI channels to assign to the track.

note ▷ For you to hear sound, the audio outputs from your hardware MIDI play-back device (your synth, sampler, drum machine or whatever) need to be connected directly to an external playback system (such as an external mixer and monitors) or they need to be routed to a playback system via the Pro Tools mixer.

tip ▷ To monitor your hardware MIDI instrument's analogue outputs, you can route these to your Pro Tools audio interface and use Auxiliary Inputs to feed the audio into the Pro Tools mixer.

note ▷ Auxiliary Inputs cannot record audio to disk – they just monitor the input and feed this into the mix. To record audio to disk you can either use the Bounce To Disk command or you can route the audio from the Auxiliary input into an Audio track using an internal Pro Tools bus and record onto this.

Assigning Multiple MIDI Output Assignments

MIDI programmers and synthesizer players often like to layer up synthesizer sounds – either before they are recorded or afterwards when working on the mix.

There are several ways to do this – such as within the synthesizer itself. Nevertheless, a convenient way can be to simply assign multiple destinations to a single MIDI track. So, for example, you might route a keyboard pad to a Reason module and to the SampleTank 2 plug-in at the same time, choosing suitable patches on each to create your layered sound.

This is easy to do in Pro Tools – just Control-click (Macintosh) or Start-click (Windows) on the track's MIDI Output Selector and use this popup selector to choose additional destinations from any of the devices and channels available on your system.

Using Virtual Instruments

With previous versions of Pro Tools, you had to use both a MIDI track and an Auxiliary or Audio track to monitor the plug-in or external MIDI device. Pro Tools 7 LE and M-Powered sessions provide up to 32 dedicated Instrument Tracks that make it much easier for you to set up external MIDI synthesizers, samplers, drum machines and suchlike for use with Pro Tools. These Instrument Tracks are also intended to be used with virtual instrument software plug-ins and ReWire.

note ▷ You can still use a separate MIDI track and an Auxiliary or Audio track with the plug-in inserted on this if you prefer. The main advantage of the Instrument track is to save you the trouble of setting two tracks up.

Instrument tracks are a kind of hybrid of MIDI and Audio tracks that provide both MIDI and audio capabilities in a single channel strip, allowing you to record MIDI and monitor audio from instrument plug-ins or external MIDI devices.

You can record and edit MIDI data on these tracks that will trigger an external MIDI device that you have hooked up, or a virtual instrument that you have inserted on the Instrument track.

You can insert signal-processing plug-ins into any (or all) of the remaining four Inserts to process audio from the virtual instrument plug-in or external MIDI device.

tip ▷ Any virtual instrument plug-ins that support audio input mode can be used to further process audio coming from the first instrument on the Instrument track by inserting these into subsequent Insert slots.

You can also record and play back MIDI controller data on the Instrument track and automate all parameters directly in the Instrument track.

tip ▷ Instrument tracks support multiple MIDI output assignments, just like regular MIDI tracks, letting you combine sounds from virtual instruments and external MIDI devices – triggering these all from the same Instrument track.

Using Instrument tracks

Step 1. Create a new Instrument track.

Step 2. Either:

Insert a virtual instrument plug-in into the Inserts section of the Instrument track (in which case the MIDI output in the track's Instruments section is automatically assigned to the inserted instrument plug-in).

Fig. 6.19 – Instrument track showing Instruments and Inserts sections with the MIDI output from the Instruments section assigned to SampleTank 2 and the SampleTank 2 plug-in instantiated in the first track insert slot.

Or:

Route the audio output from an external MIDI device such as a synthesizer or sampler into the Instrument track's audio input then set the MIDI output in the track's Instruments section at the top of the Channel strip to the appropriate MIDI port and channel for the external MIDI device.

Step 3. Record-enable the Instrument track and play your MIDI controller. You can now record MIDI data to the Instrument track while playing the instrument plug-in or external MIDI device.

Fig. 6.20 – This Instrument Channel sends MIDI to a DX7 and monitors audio from the DX7 via Analog Inputs 7/8 on the Pro Tools interface – to which the DX7's audio outputs are connected in this example.

Using Reason, Live, and SampleTank

Every Pro Tools system comes with a bundle of extra software that includes Propellerhead Software Reason Adapted, Ableton Live Digidesign Edition, and IK Multimedia SampleTank SE – each of which is an extremely powerful software application in its own right.

Reason Adapted gives you a virtual rack of interesting MIDI-based synthesizers, samplers, drum machines and effects modules, and also has its own sequencers. SampleTank SE is just one sample playback module – but this has an excellent library of sounds to suit most kinds of music. Ableton Live is an incredibly flexible sequencer with its own sampled and synthesized sounds that can be used to create, modify and playback loops, phrases and songs using its 'elastic audio' technology. Live also allows you to import samples and loops from other libraries. Change the tempo, and the loops change tempo automatically. Choose a new loop running at a particular tempo, and this will automatically run at the tempo of your sequence.

SampleTank works as a virtual instrument plug-in while Reason and Live run as stand-alone applications that connect to Pro Tools using Propellerhead's ReWire technology. So you can use Reason or Live for loop- and sample-based composition, and have the audio from both piped directly into your Pro Tools sessions. You can even use both at the same time within the same Pro Tools Session – although, predictably, this works best on faster CPUs with plenty of RAM (random access memory).

ReWire

ReWire was developed by Propellerhead Software to make it possible to transfer MIDI and audio data between applications running on the same computer without using any external connections – ReWire makes 'virtual' connections internally.

Currently there are two main applications that use ReWire with Pro Tools – Propellerhead Reason and Ableton Live. ReWire support is also being developed by other third-party companies.

So how does ReWire let you route audio into Pro Tools? Well, compatible ReWire client applications are automatically detected by Pro Tools and Real-Time AudioSuite (RTAS) plug-ins for these are added to the list of available plug-ins. These ReWire plug-ins then allow you to route audio coming from the ReWire 'client' application into Pro Tools mixer channels.

note ▶ When you insert a ReWire plug-in into the Pro Tools mixer, Pro Tools automatically launches the corresponding ReWire 'client' application if the client application supports this feature. If not, you must launch the ReWire application manually.

And how does the MIDI work? It couldn't be easier for the user: when the ReWire 'client' application has been launched, the MIDI inputs for that application automatically become available as destinations in the MIDI Track Output selectors in Pro Tools.

The timing of the linked applications (Pro Tools and the ReWire client) is synchronized with sample accuracy and you can use the transport controls on either application to control the other. Pro Tools transmits both Tempo and Meter data to the ReWire client application, allowing the ReWire application's sequencers to follow any tempo and meter changes in the Pro Tools session.

For example, you may have recorded sequences using Reason that you want to play back in sync with your Pro Tools session.

note ▷ With the Pro Tools Conductor button selected, Pro Tools always acts as the Tempo master, using the tempo map defined in its Tempo Ruler. With the Pro Tools Conductor button deselected, the ReWire client acts as the Tempo master. In both cases, playback can be started or stopped in either application.

Of course, once the audio outputs from Reason or Live are routed into Pro Tools you can process these incoming audio signals with plug-ins, automate volume, pan, and plug-in controls, and use the Bounce To Disk command to 'fix' the incoming audio as files on disk.

How many channels of audio does ReWire support with Pro Tools LE? In theory, up to 64 are available, but, in practice, performance is determined by several factors – including host CPU speed, available memory, and buffer settings. The manual states that Digidesign cannot guarantee 64 simultaneous audio channel outputs with ReWire on all computer configurations and also points out that 48 kHz ReWire applications cannot be used in 96 kHz Pro Tools sessions.

note ▷ Some ReWire client applications, such as Reason, support a single stereo output path (Reason's remaining 62 outputs are mono only). If you want to use multiple stereo outputs with Reason you will need to insert multiple multi-mono ReWire plug-ins, then unlink and assign left and right outputs separately using the Link Enable and Channel selector buttons in the plug-in's window.

Troubleshooting

If you are using ReWire, especially on a single-processor computer, the load on the CPU can easily become too great, especially if you are using several RTAS plug-ins in your Pro Tools session. CPU overloading causes all kinds of problems including distorted audio and other performance errors.

For example, if you try to create or open a session with more plug-ins than the host CPU can support, Pro Tools LE may repeatedly lose communication with the Digi 002 or Digi 002 Rack and report a −6097 error, or Pro Tools may freeze.

To recover from this condition, you may have to turn off the Digi 002 or Digi 002 Rack and quit Pro Tools when prompted. Be sure to reduce the number of plug-ins before playing back the session on the same system.

Another common error that you may encounter is Error −6093, 'Operating System Held Off Interrupts for Too Long'. If this happens a lot, try increasing the Hardware Buffer Size.

Looping Playback when using ReWire

You can set a playback loop either in Pro Tools or in the ReWire client.

If you want to loop playback in Pro Tools, just click and hold the mouse button and drag across the ruler in the Edit window to select the time range that you want to loop in the Pro Tools timeline before starting playback.

Or you can set loop or playback markers in the ReWire client sequencer before you start playback.

note ▷ If you create a playback loop by making a selection in the Pro Tools timeline, once playback is started, any changes made to loop or playback markers within the ReWire client application will deselect the Pro Tools Timeline selection and remove the loop.

Also, some ReWire client applications, such as Reason, may misinterpret Pro Tools meter changes, resulting in mismatched locate points and other unexpected behaviour. So you should avoid using meter changes in Pro Tools when you are using Reason as a ReWire client.

Propellerhead Reason

Overview

Reason is a collection of virtual instruments and audio processors that can be assembled in a window that looks like a rack of MIDI and audio hardware. You can choose what to put into this 'rack' from a menu that contains a range of MIDI synthesizers, samplers, pattern sequencers, drum machines, effects devices and a mixer. Reason has its own main sequencer that, although quite easy to learn, can be 'fiddly' to use – and its features are no match for those in Pro Tools.

All Reason's rack devices operate similarly to their hardware equivalents. For example, the mixer closely resembles the Mackie 3204 rackmount model with its fourteen stereo channels, 2-band EQ, and four effects sends. Sound modules include the Subtractor and Malstrom synthesizers, NN19 and NN-XT sampler players, Dr Rex Loop Player and Redrum – a TR808-style drum machine. Reason also provides a wide range of effects including reverbs, delays – and even a vocoder.

A recently added feature is called the Combinator. This rack module can contain a selection of Reason devices along with associated effects and modulation routings. You can save these Combinator 'patches' as 'Combis' so that you can instantly recall any setup, whether a complex chain of effects or a split or layered multi-instrument. For example, the MClass Mastering Suite is a ready-made Combi containing a suite of new mastering tools – a 4-band EQ, a Stereo Imager, a Compressor with side-chain and soft-knee options, and a Maximizer to make your tracks louder.

Fig. 6.21 – Reason virtual 'rack' showing its MClass Mastering Suite of processors.

Reason has become extremely successful as a suite of affordable, popular virtual instruments that can be used stand-alone to develop instrumental arrangements that are then incorporated into more ambitious projects in Pro Tools where vocals and live instruments can be added and additional virtual instruments and plug-ins can be used to complete the production.

Recording Reason synths

Step 1. To use Reason with Pro Tools, you must launch Pro Tools first and open a new or existing session.

Step 2. You then add a MIDI Track that you will use to play back one of Reason's rack of synthesizers and an Auxiliary Input or Audio Track to receive the audio output from this synthesizer – or add an Instrument Track which handles the MIDI and the audio monitoring on the one track.

Step 3. In the Mix window, insert the RTAS ReWire plug-in for Reason using the Audio or Instrument track's Inserts section. Reason will launch its Default Song automatically in the background as soon as you insert the ReWire plug-in.

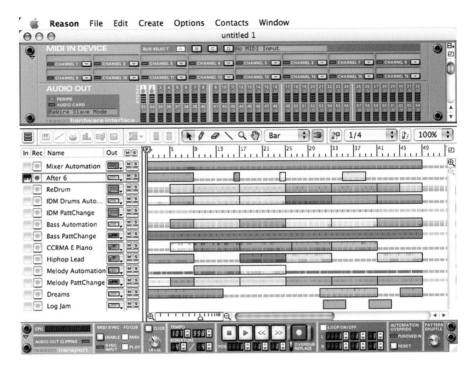

Fig. 6.22 – Reason Default Song.

Step 4. The ReWire plug-in's window will also open – ready for you to choose the outputs that you want to use from Reason. Select the outputs that you want to use by clicking on the popup list located at the lower right of the plug-in's window.

Fig. 6.23 – Reason ReWire Plug-in with Reason's Mix L and Mix R output pair selected.

Setting up the Reason Synthesizers

Normally, when you launch Reason, the standard Default Song document opens. This contains a mixer, various effects modules and several Reason synthesizer modules – along with the demo song sequences.

If you are using Reason with Pro Tools, it often makes more sense to avoid using Reason's mixer and effects – because you can use the much more full-featured mixer and higher quality effects in Pro Tools itself.

In this case you can prepare a custom default song containing just the Reason devices you plan to use, with their outputs connected directly to the Reason hardware interface. This hardware interface then feeds Reason's audio outputs directly into Pro Tools via the ReWire RTAS plug-ins.

Step 1. To set this up, you should set the General Preferences in Reason to open an Empty Rack first.

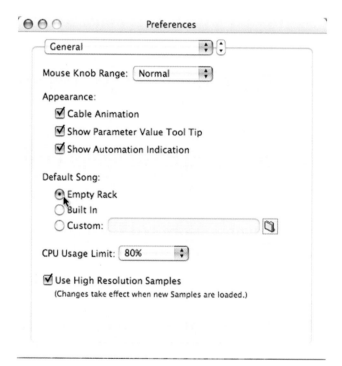

Fig. 6.24 –
Reason General
Preferences.

Step 2. Close the Default Song, if it is open, and select 'New' from Reason's File menu to open a new empty Reason rack.

Step 3. Add a selection of devices to the rack using Reason's Create menu. Each time you create a new device in your Reason rack, a sequencer track with its output set to this device will be automatically added to Reason's main sequencer.

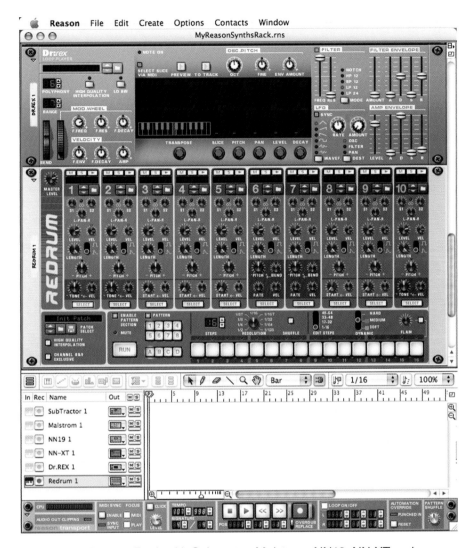

Fig. 6.25 – Reason Rack with Subtractor, Malstrom, NN19, NN-XT and Redrum modules added and with the Sequencer window showing empty tracks for these.

tip ▶ If you are not intending to use Reason's main sequencer, you can click and drag on the dividing line between the sequencer and the modules above it until it is completely hidden from view. You can also collapse Reason's Transport controls by clicking the small arrowhead at the far left of this module.

Step 4. Press the Tab key on your computer keyboard to reveal the back of the Reason rack. Here you can choose which modules are connected to which outputs.

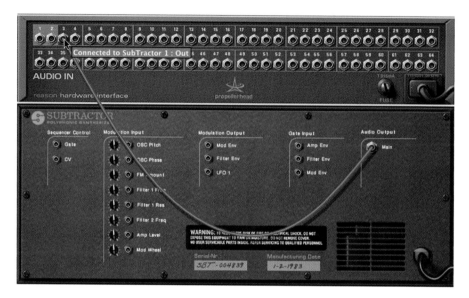

Fig. 6.26 – Rear of Reason Rack showing module output connections.

Step 5. Connect each module in turn to individual Reason outputs. You can then route these to individual Pro Tools mixer inputs using ReWire.

tip ▷

You may prefer to mix the audio outputs of the various Reason modules inside Reason using its internal 14:2 mixer. If so, add this module to your Reason rack, before connecting any modules to outputs, and connect the Reason modules' outputs to this mixer instead.

Then connect the 14:2 mixer outputs to, say, Reason outputs 1 and 2, and route these to a stereo Pro Tools track input using ReWire. This way, you can audition all your Reason modules while setting them up and choosing their sounds without having to switch tracks in Pro Tools: just choose a different Reason module as the MIDI destination each time you want to listen to this.

Fig. 6.27 – Module outputs routed to the Reason 14:2 mixer which is routed to Reason outputs 1 and 2.

Step 6. If you will not be using the Reason sequencer, then it makes a lot of sense to hide the sequencer section and the transport controls in Reason.

Fig. 6.28 – Reason Rack containing 14:2 mixer and 6 sound modules and with the sequencer and transport controls hidden.

Step 7. When you have set up your Reason rack the way you want it, you should save this into your Reason folder. You might call this 'MyReasonSynths' – or whatever works for you.

Step 8. Now change Reason's General Preferences to open a Custom song as default, and select the 'MyReasonSynths' song document by clicking on the small Folder icon at the right of the Prefs window and navigating through your disk drives and folders until you find the 'MyReasonSynths' song document in whichever folder you saved it.

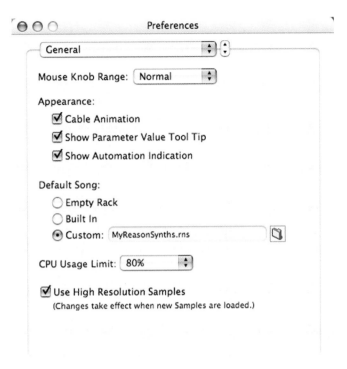

Fig. 6.29 – Reason's Preferences set to open a Custom Default Song.

note ▷ Once your custom default song is set up in Reason, this is what will be opened each time you launch Reason or open a New document from Reason's File menu.

Reason uses the Default Song file as a template on which to base any new Reason document, so the document opens as 'Untitled'.

Reason expects that you will record new sequences into its own sequencers and that you will subsequently save this new document (based on the default song template) as a new file – named 'My New Song' or whatever you want to call your new song.

If you do not record any new sequences into Reason or make any changes to the configuration of the rack modules, then there will be no need to save a new Reason document before ending your Pro Tools Session. Next time you open this Pro Tools Session, the same 'Untitled' Reason document based on the Default Song will open, so, unless you have changed the Reason Default Song in the meantime, the same set of Reason modules will be available.

Of course, if you have selected different patches for your synthesizers, added or removed any modules to or from your Reason rack, or made any changes to the module settings that you want to keep, then you will need to save a copy of this edited Reason song document – preferably into the Pro Tools folder for your session so that it can be backed up (hint, hint!) with the Session.

Routing MIDI from Pro Tools to Reason

When you have inserted a Reason RTAS plug-in and Reason has launched, all the available Reason sound modules in the current Reason document will be listed as Pro Tools MIDI Track Output destinations.

Step 1. Choose a rack module from among the Reason devices listed as MIDI Output destinations.

Dr.REX 1 - channel-6
Hardware Interface - channel-1
Kontakt2 1 ▶
Malstrom 1 - channel-3
✓ NN19 1 - channel-4
NN-XT 1 - channel-5
Oxygen 8 - (No Channels Activated)
Redrum 1 - channel-7
SampleTank 2 1 ▶
SubTractor 1 - channel-2

Predefined ▶

Fig. 6.30 – Pro Tools MIDI Output destinations with Reason Modules available.

Step 2. To play the Reason sound modules from a MIDI keyboard connected to your MIDI interface, all you need to do is to make sure that MIDI Thru is selected in the Pro Tools Options menu and that the Record button is enabled in the Pro Tools MIDI track. Now you are all set up and ready to record using the Reason module you have chosen.

Fig. 6.31 – Reason Instrument Track.

Step 3. Go ahead and record some stuff for your Session: press Record then Play in the transport window to start recording and press the spacebar to stop.

Recording from Reason into Pro Tools

When you are satisfied with your Reason recording, you should either bounce it to disk or record directly to Pro Tools audio tracks in real time. Though you can hear the bounce being created in real time, you cannot adjust the mixer or other controls during a Bounce to Disk, so I recommend that you record to tracks instead.

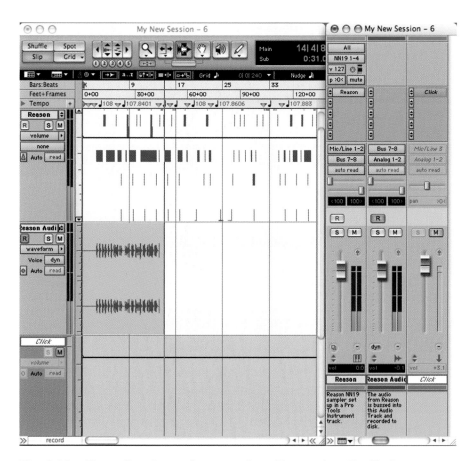

Fig. 6.32 – Recording the audio output from Reason into Pro Tools.

Step 1. Use the Track menu to create a new stereo Audio track.

Step 2. Choose an unused stereo bus pair as its input.

Step 3. Change the audio output of the Instrument track that is being used to monitor Reason to the same stereo bus pair so that the output of the Instrument track is routed to the input of the Audio track.

Step 4. Record-enable the Audio track.

Step 5. Make sure that the counter is at the location you wish to record from (e.g. press Return if you want to start from the top) and start recording by pressing 'Play' then 'Record' in the Transport window.

Step 6. Hit the spacebar to stop recording.

Developing and recording material using Ableton Live

Overview

Ableton's Live is one of those 'must-have' software applications for putting beats and basslines together with MIDI – building arrangements up with pads, bells and whistles (or whatever) and dropping in samples all over the place until you hear something you like. Live lets you quickly rearrange the order in which sections of your music play and makes it simple to play parts from one section alongside another section. And when you drop in sampled loops, these automatically play at the correct tempo. All amazingly helpful!

Live's Arrangement View is used for multitrack recording and editing while the Session View lets you do the 'jamming' – improvising the arrangement on the fly.

Programming beats into Live is very easy:

Step 1. Open the Impulse drum samples folder from the Devices Browser and drag a preset kit into a MIDI track in Live's Session View.

Step 2. Click on the Arm button so it can receive MIDI, then play your MIDI keyboard or hit the middle row of keys on your computer keyboard to play the drum sounds.

Step 3. Double-click any empty slot in the Session view to create a MIDI clip to record into.

Step 4. Click the clip's Play button to activate recording – then start dancing on those keys!

One of the most impressive things about Live is the speed with which you can do things. Say you have programmed a 1-bar pattern and you want to change this to a 4-bar pattern. Just click and drag on the Loop Length parameter and it instantly changes to the new length – even while the pattern is playing back. So you can keep on building up patterns while listening to your music build up before your very ears!

Recording basslines and leadlines is equally simple using the built-in sampler instrument – called Simpler. This has enough bass sounds, pads, choirs, assorted keyboards, and strings to let you put relatively ambitious musical arrangements together on top of your beats. As with Impulse, you can play this from the computer keyboard. This has an octave of notes laid out on the middle row of keys with the sharps and flat on the upper row of keys. You can transpose the range by hitting the Z or X keys and change the velocity of the notes using the C and V keys. So you can do lots of stuff with Live just using a laptop – without a MIDI keyboard!

A third built-in instrument, Operator, is a powerful synthesizer that, like Impulse and Simpler, is totally integrated such that every parameter can be automated or performed in real time. Live also has lots of built-in effects such as Phaser, Flanger, Arpeggiator and my favourite – Beat Repeat. This lets you create short loops on the fly, controlling their lengths manually or via random functions for endless variations. And there's more: support for MP3 and other compressed formats means that you can now use just about any audio file in Live. Like Reason's Combinator patches, Live's Device Groups feature allows you to save Simpler, Impulse and Operator instruments with chains of MIDI and audio effects attached as presets.

These features have all proven to be extremely popular with DJ's and remix-ers who appreciate the way that Live lets you make spontaneous creative decisions while building up ideas.

Using Ableton Live with Pro Tools

One way to use Live with Pro Tools is to build sequences and arrangements using Live stand-alone, then open these for playback into the Pro Tools mixer. With this setup, you can conveniently record the output from Live onto Pro Tools tracks, then quit Live to reduce the load on your CPU. With the Live material in Pro Tools, you can then the record additional material alongside this.

Another way to work is to build sequences and arrangements in Live while it is being hosted by Pro Tools. Within this scenario you can always record material directly into Pro Tools at any stage during the session. However, this is a very CPU-intensive way to work, so it will only work well with the very fastest systems.

Step 1. To use Live with Pro Tools, launch Pro Tools first and open a Session.

Step 2. Create an Auxiliary track to monitor Live's audio output.

note ▷ You could use an Audio track to monitor Live, but this uses a playback 'voice' and you only have 32 of these available. Also, if you want to record the audio from Live into your Pro Tools Session, then you will need to use a separate Audio track as well as the monitor track, as Pro Tools won't allow you to record into a track being used to monitor a plug-in.

Step 3. Insert an Ableton Live ReWire plug-in into the audio track's Inserts section.

Step 4. Choose the Live outputs that you wish to use from the popup selec-tor at the lower right of the ReWire plug-in window. The Mix L–Mix R pair is the obvious choice as default.

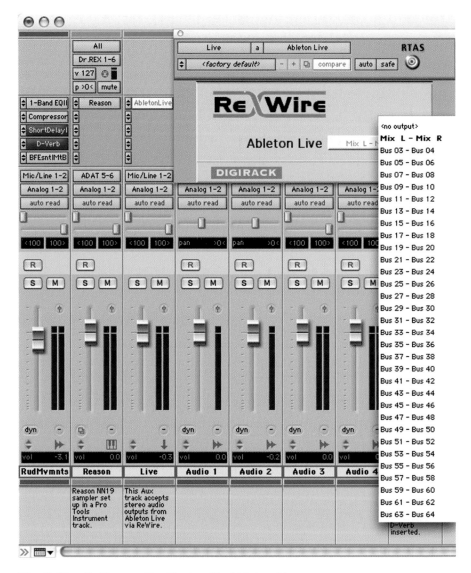

Fig. 6.33 – Setting up Pro Tools with Ableton Live.

Step 5. Unlike Reason, Live does not launch automatically in the background as soon as you insert the ReWire plug-in, so the next step is to launch this manually. When Live launches (while Pro Tools is running with an Ableton Live ReWire plug-in inserted), it automatically connects to Pro Tools via ReWire.

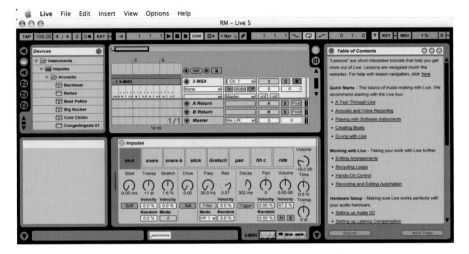

Fig. 6.34 – Ableton Live ReWired into Pro Tools.

Step 6. Now you can press Play either in Pro Tools or in Live, and both applications will start to play back in sync.

At this point you could play back an existing Live sequence that you recorded earlier into Live or that you obtained from someone else, and play this in sync with your Pro Tools Session. Or you could open a new Live sequence and record new material directly into Live.

Live has its own internal mixing features which you can use while building your arrangement in Live and you can balance the audio coming via ReWire from Live with the audio from the rest of your Pro Tools tracks using the Auxiliary Input fader.

Recording from Ableton Live into Pro Tools

When you have the sequences that you want playing back in Ableton Live, you should convert these into audio on separate Audio Tracks in Pro Tools. It is possible to bounce audio playing through Auxiliary Inputs to disk by selecting and soloing the audio and choosing the Bounce To Disk command from the File Menu. Alternatively, you can record directly to Pro Tools audio tracks in real time.

Recording the audio output from Ableton Live into Pro Tools works the same way as recording audio from any virtual instruments into Pro Tools: The Auxiliary Input that you are using to monitor the audio output from Live will not allow you to record this audio to disk. The solution is to route the audio from the Auxiliary Input into an Audio track using the internal buses.

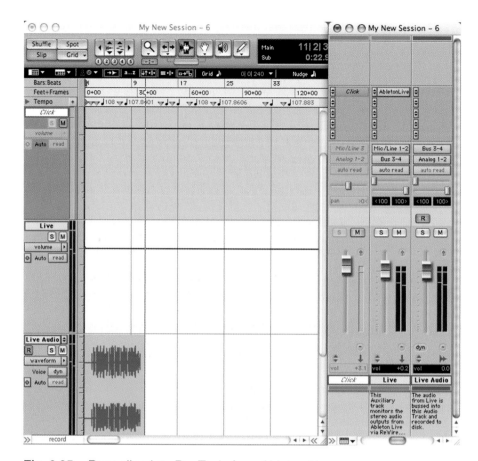

Fig. 6.35 – Recording into Pro Tools from Ableton Live.

Step 1. Use the Track menu to create a new stereo Audio track.

Step 2. Choose an unused stereo bus pair as its input – Bus pair 3–4 in this example.

Step 3. Change the output of the Auxiliary track that is being used to monitor Live to the same stereo bus pair, so that the output of this Auxiliary track is routed to the input of the Audio track.

Step 4. Record-enable the Audio track.

Step 5. Make sure that the counter is at the location you wish to record from (e.g. press Return if you want to start from the top) and start recording by pressing 'Play' then 'Record' in the Transport window.

Step 6. Hit the spacebar to stop recording.

IK Multimedia SampleTank 2

Overview

IK Multimedia's SampleTank 2 is a 16-part multitimbral sample playback plug-in that comes with a large library of samples to get you started. You can also import your own WAV, Audio Interchange File Format (AIFF) or Sound Designer II (SDII) files and it can read Akai and SampleCell formats directly. SampleTank 2 offers 32-bit and 96 kHz quality playback with up to 256-note polyphony – depending on your hardware. It is supplied in two versions – L and XL. The sample library in the L version comes on 4 CD-ROMs while the XL version comes on 8 CD-ROMs and includes more than 1500 sounds, encompassing loops for house and techno, a good selection of electronic and acoustic drums, percussion, ethnic instruments, choirs and voices, synths, guitars, basses, strings, brass, woodwinds, and keyboards – just about everything you might need to put music together.

SampleTank 2 features three synthesizer 'engines' that provide much more flexibility than most of its competitors. Besides conventional sample play-back you also get a pitchshift/timestretch mode that lets you independently control pitch and tempo for loops, like ACID or Live. And the SampleTank Time Resynthesis Technology (STRETCH) engine provides even more control over the tempo, pitch and harmonics of the samples, so when you transpose a sample into a different musical key, or run a loop at a different tempo, it still sounds natural. Stretch mode preserves formants as you shift pitch, offering enhanced realism while bending notes and gliding or playing polyphonic chords with phrases. So you can bend a singer's voice using the pitch wheel on your synth and the voice will follow without the chipmunk effect. And if you play chords, the notes within the chord will all end at the same time – unlike with a conventional sampler.

The interface is very straightforward, which makes it easy and quick to use, and it has everything you need in one compact window. All the available library instruments are listed in the centre of the interface window, which functions as a browser. I am always impressed at how fast the sounds load – just double-click on any instrument and up it comes in the selected multitimbral part in the Mix section to the left of the browser, ready for you to play almost instantaneously.

The Mix area lets you control the 16 multitimbral parts. Here you can set the MIDI channels, switch the solo and mute functions, set the volume and pan controls, and choose the polyphony for each part. I particularly like the 'one-click' Loop Sync feature that lets you sync any loop to your sequencer's tempo with a single click – it couldn't get much easier than this! The MIDI control features are also excellent. Any of SampleTank 2's controls can be mapped to any MIDI source and any MIDI program change message can be mapped to any SampleTank 2 sounds for instant recall.

Typical synthesizer controls – filters, LFOs and envelopes – can be used to create new sounds using the library samples as oscillator waveforms – ideal for creative sound designers. A popup selector lets you switch each instrument between polyphonic, monophonic, and monophonic legato playback modes. A special

Macro controls section is also provided. This has up to four controls that control different parameters depending on which instrument they are associated with.

The Effects section provides 32 DSP effects and up to five of these can be used with each instrument. You get all the usual suspects here, including Reverb, Ambience, Delay, Filter, Wah-Wah, Chorus, AM and FM Modulation, Flanger, Autopan, Tremolo, Rotary Speaker, Lo-Fi and Distortion, along with the EQ, Compressor and Limiter from T-RackS and the Preamp, Tone Control and Cabinet simulations from AmpliTube.

A mini keyboard displays the notes being played and lets you play the sounds using the mouse. You can even play chords at the click of a mouse – selecting the chord type from the popup to the right of the keyboard. A click on the Zone button above this reveals how the samples are mapped on the keyboard using different colours.

So what are the sounds like? Well, the TR808 samples are the best I have heard yet – and I have a real TR808 to compare with! The acoustic drum kits all have long-decaying cymbals, full-sounding toms and very useable sounding bass and snare drums – and the Studio kit does sound exactly like a kit played in a small, dead studio! The DX piano sounds just perfect and the acoustic grand piano is very useable. The Pop Violins sounds just like the cheap string synthesizer that I used in the 70s and the B3 organ sounds very much like a B3 – and, again, I have a real Hammond to compare with. There is a great percussion selection as well, with ethnic instruments such as the darbuka and dumbeck, triangle, tambourine, shakers – and great congas. These congas sound very realistic and you get several different types of slaps and hits. On the other hand, I found the orchestral samples rather disappointing. The solo strings and woodwinds were far too noisy for my liking – although they were of comparable quality to similarly priced sample libraries.

Fig. 6.36 – SampleTank 2.

All-in-all, SampleTank 2 is extremely easy to use, the samples load like lightning, and the supplied library sounds very good.

Adding sampled material using SampleTank

Step 1.   Add a new Instrument track from the Track menu.

Step 2.   Insert a SampleTank 2 plug-in into the first Insert in the Instrument track. The MIDI output in the Instruments section at the top of the mixer channel strip will automatically be routed to MIDI channel 1 on this instance of the SampleTank 2 plug-in.

Step 3.   Choose a MIDI input device using the MIDI Input popup selector in the Instruments section at the top of the channel strip in the Pro Tools mixer. This defaults to 'Any', so you can skip this step if 'Any' is OK for your setup.

Step 4.   Open the SampleTank plug-in, load the sound or sounds that you want to use, tweak and add effects as necessary.

Step 5.   Record-enable the track then start recording by pressing 'Play' then 'Record' in the Transport window.

Step 6.   Play your MIDI keyboard or other MIDI device.

Step 7.   When you have finished recording, click Stop in the Transport window, or press the Spacebar. The newly recorded MIDI data will appear as a MIDI region on the track in the Edit window and in the Regions List.

Recording from SampleTank into Pro Tools

When you are satisfied with your SampleTank recording, you should either bounce it to disk or record directly to Pro Tools audio tracks in real time. Though you can hear the bounce being created in real time, you cannot adjust the mixer or other controls during a Bounce to Disk, so I recommend that you record to tracks instead.

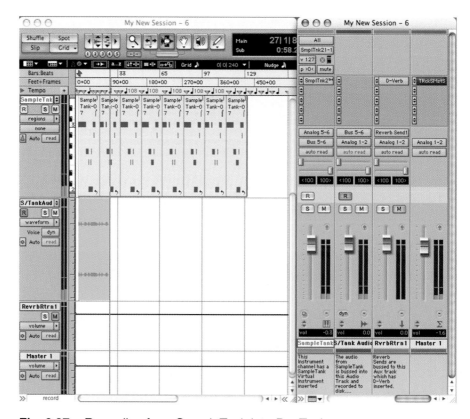

Fig. 6.37 – Recording from SampleTank into Pro Tools.

Step 1. Use the Track menu to create a new stereo Audio track.

Step 2. Choose an unused stereo bus pair as its input.

Step 3. Change the audio output of the Instrument track that is being used to monitor SampleTank to the same stereo bus pair so that the output of the Instrument track is routed to the input of the Audio track.

Step 4. Record-enable the Audio track.

Step 5. Make sure that the counter is at the location you wish to record from (e.g. press Return if you want to start from the top) and start recording by pressing 'Play' then 'Record' in the Transport window.

Step 6. Hit the spacebar to stop recording.

Making tracks inactive to conserve processing resources

When you have finished recording your Instrument tracks and any MIDI/ Auxiliary tracks, I strongly recommend that you should record these to disk as audio as soon as possible. Just for the record, Digidesign also recommends this.

When you have recorded these tracks as audio, you can inactivate the Instrument, Auxiliary or Audio tracks that are hosting the virtual instruments and any other plug-ins – and listen to the recorded audio tracks instead. Playing back audio tracks makes much less demands on the computer's CPU than playing back virtual instruments. Also, inactive tracks and plug-ins use no voices or digital signal-processing resources.

note ▷ Audio, Auxiliary Input, Master Fader, and Instrument tracks can be made inactive. MIDI tracks cannot.

To make a track inactive, Command-Control-click (Macintosh) or Control-Start-click (Windows) on the Track Type indicator in the Mix window. It is immediately obvious if a track is inactive, as the Playlists for inactive tracks become dimmed and track controls become greyed out. Plug-ins, sends, voices, and automation on inactive tracks are all disabled.

note ▷ Tracks may also be automatically made inactive if a session is opened on a system with less DSP power than the system that it was created on.

Recording virtual instruments as Audio

It is a good idea to record your virtual instruments as audio as soon as possible in the production process. There are two reasons for this. Firstly, you can then remove the virtual instrument so that it is no longer using DSP resources. Secondly, and possibly even more importantly, if you take your Pro Tools session to another studio that doesn't have the particular virtual instrument that you used, you would not be able to re-create the original sound that you had – but if you have recorded the audio, then you have it!

You may think that the way to record a virtual instrument is to simply use an Audio track instead of an Auxiliary track to monitor your virtual instrument – so that you can just put this Audio track into record while playing back the MIDI into the virtual instrument. Unfortunately, Pro Tools does not make it as simple as this. If you try this, you will soon discover that the audio from the virtual instrument cannot be recorded onto the same track that is monitoring the virtual instrument.

The solution is to route the output from the Auxiliary, Instrument or Audio track used to monitor the virtual instrument to the input of a separate Audio track using an internal bus – then use this separate Audio track to record the audio playing back through the Auxiliary, Instrument or Audio track.

What this chapter has explained

You should now be aware how to set up Pro Tools for use with your MIDI devices and virtual instruments, and how to record and play back MIDI and Instrument tracks.

You should also have a good understanding of how to use Reason, Live and SampleTank with Pro Tools – and what these can be used for.

You should understand why it is a good idea to record your MIDI and virtual instruments as audio tracks as soon as possible in the production process and how to do this.

You should now be in a position to work on your music production using the virtual instruments supplied with Pro Tools LE to at least 'map out' the arrangement before you start recording real instruments to use alongside or to replace the virtual instruments.

And if you are making electronic music, Reason, Live, and SampleTank may be all you need to complete your production.

7

Editing MIDI

Pro Tools does not have as many advanced MIDI editing features as Logic or Cubase. Nevertheless, all the basics are there, along with several neatly implemented features that make Pro Tools a joy to work with compared with some of its more complicated competitors.

Editing Basics

Pro Tools lets you edit MIDI data alphanumerically or graphically, according to your preference. You can use the standard Cut, Copy, Paste, and Clear commands to manipulate MIDI data, and Pro Tools also has 'Special' versions of these commands that provide extra functions for MIDI data.

There are many useful keyboard commands and display zooming features that it will pay you to become familiar with, and the recently added MIDI Properties feature lets you apply changes to regions or tracks with a minimum of fuss.

The MIDI Operations windows let you transform MIDI data in a variety of ways, so time spent learning its wealth of features will be well rewarded.

Bars, Beats, and Clock 'Ticks'

When you are working on music, you will normally want to display the Bars: Beats ruler so that you can edit according to the bar lines and correct note positions.

MIDI data is recorded into Pro Tools with a very high degree of accuracy. The internal 'clock' to which MIDI is resolved has an incredible 960,000 pulses per quarter note (PPQN) resolution. However, when the Time Scale is set to Bars:Beats, the *display* resolution in Pro Tools is 960 PPQN, which provides more manageable numbers to work with.

When you are editing MIDI data using Bars and Beats, there are several sets of circumstances where you may want specify tick values. For example, when placing and spotting regions; when setting lengths for regions or MIDI notes; when locating and setting play and record ranges (including pre- and post-roll); when specifying settings in the Quantize and Change Duration windows; and when setting the Grid and Nudge values.

Bars are sub-divided into beats, and beats are sub-divided into 'ticks', so the locations of MIDI notes are specified in terms of bars, beats, and ticks. Take a look at the accompanying screenshot to see a selected MIDI note with its location and length displayed in the Event Edit area above and in the MIDI Event List to the right. The note starts at bar 4, beat 1, tick 591 and the note lasts for 724 ticks, so it ends at bar 4, beat 2, tick 355.

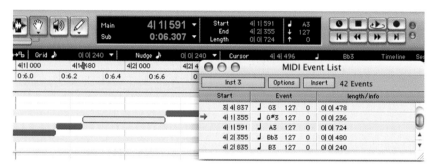

Fig. 7.1 – A MIDI note displayed graphically with its parameters displayed in the Event Edit area above and in the MIDI Event List to the right.

It helps to become familiar with the numbers of ticks that correspond to the main normal, dotted and triplet note lengths. For a half note these are 1920, 2880, and 1280 ticks. For a quarter note they are 960, 1440, and 640 ticks. For an 8th note they are 480, 720, and 320 ticks. For a 16th note they are 240, 360, and 160 ticks. For a 32nd note they are 120, 180, and 80 ticks, and for a 64th note they are 60, 90, and 40 ticks.

Graphic Editing

Pro Tools MIDI tracks can be edited graphically in the Edit window. Here you can use the standard Pro Tools Trimmer tool to make notes shorter or longer and use the Grabber tool to move the pitch or position – or 'draw' notes in using the Pencil tool. You can draw in or edit existing velocity, volume, pan, mute, pitchbend, aftertouch and any continuous controller data, and the Pencil tool can be set to draw freehand or to automatically draw straight lines, triangles, squares, or randomly. The Pencil tool also lets you draw and trim MIDI note and controller data and the Trim tool can trim MIDI note durations when a MIDI track is set to Velocity view.

It can be very handy at times to insert notes using the pencil tool instead of setting up an external keyboard. Just make sure that the MIDI track is in Notes view and select the Pencil tool at the top of the Edit window. To insert quarter notes on the beat, for example, set the Time Scale to Bars and Beats, then set the Edit mode to Grid and the Grid value to quarter notes. As you move the Pencil tool vertically and horizontally within a MIDI track, the pitch and the bar/beat/clock location is displayed just above the rulers in the Edit window. When you find the note and position you want, you can just click in the track to insert it. It's as simple as that!

To select notes for editing, you can use either the Grabber tool to drag a marquee around the notes, or drag using the Selector tool across a range of notes. Once some notes are selected, you can drag these up or down to change the pitch using the Grabber or Pencil tools – while pressing the Shift key if you want to make sure that you don't inadvertently move the position of the notes in the bar. You can use the Trimmer tool or the Pencil tool to adjust the start and end points of the notes. If you set the MIDI track to Velocity view, you will see the attack velocities of the notes represented by 'stalks'. You can edit these using the Grabber tool.

Sometimes, you may simply need to edit one note. If you select a note using the Grabber or the Pencil, its attributes will be displayed in the Event Edit area to the right of the counters at the top of the Edit window. Here you can type in new values for any of the displayed parameters. See the accompanying screenshot for more details.

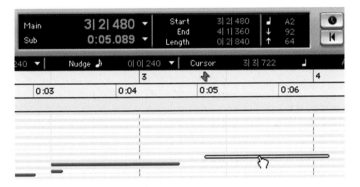

Fig. 7.2 – A selected MIDI note's attributes are displayed in the Event Edit area to the right of the Counters, ready for editing.

Continuous controller data recorded onto MIDI tracks is displayed as a graph line with a series of editable breakpoints. These breakpoints are stepped to represent individual controller events, in contrast to the standard automation breakpoints, which are interruptions on a continuous line.

You can edit pitchbend, aftertouch, mod wheel and other MIDI controller data directly in the Edit window according to which type of data you select using the popup Track View selector.

Fig. 7.3 – MIDI Track View selector.

For example, you can use the Pencil tool to draw in pitch bend data either to edit a played performance or to create a new pitch bend 'from scratch'.

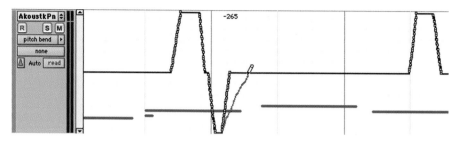

Fig. 7.4 – Editing MIDI Pitch Bend data.

note ▷   MIDI Controller #7 (Volume) and #10 (Pan) are treated as standard automation data and can be recorded and automated using the volume and pan controls in the Mix window. These can also be recorded from external MIDI devices.

Event List Editing

You can also view and edit MIDI data using the MIDI Event List, which is available from the Event menu. This lists MIDI events alphanumerically with letter names for the notes and numbers for the On and Off velocities, locations and lengths. In this window, you can select, copy, paste, or delete events and edit values by typing directly in the list.

Start	Event			length / info
1| 4| 160	E2	80	64	0| 2| 463
2| 3| 000	G2	64	64	0| 2| 900
2| 3| 000	F2	72	64	0| 0| 269
3| 2| 480	A2	92	64	0| 2| 840
4| 1| 800	G2	71	64	0| 2| 874
8| 4| 013	E4	80	64	0| 2| 290
9| 1| 925	E5	80	64	0| 0| 615
9| 2| 453	E4	72	64	0| 0| 618
9| 2| 926	E5	80	64	0| 0| 822
9| 3| 685	E4	64	64	0| 0| 947
9| 4| 470	E5	71	64	0| 0| 946

MIDI Event List — Akoustik Piano — Options — Insert — 467 Events

Fig. 7.5 – Data in the MIDI Event List window.

There are three popup selectors at the top of the MIDI Event List window. The first lets you choose which track to display in the MIDI Event List window.

The second popup, labeled Options, has various commands that let you customize the MIDI Event List, make it scroll the way you want it to, and choose where events are to be inserted. You can also choose what to display in the list using the View Filter.

The third popup, labeled Insert, lets you insert any MIDI event (except SysEx) into the list.

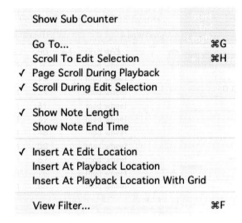

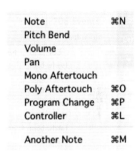

Fig. 7.6 – MIDI Event List Options popup selector.

Fig. 7.7 – MIDI Event List Insert popup selector.

note ▶ There is no list editor for audio data – a major omission in my opinion.

Remove Duplicate Notes

One of the most basic, yet useful, MIDI editing commands in Pro Tools 7 is 'Remove Duplicate Notes'. It is very easy to accidentally hit a note on a MIDI keyboard twice – you might hit it tentatively just before it should be played and then again more positively on the beat, or a little late.

If a note starts within the first 25% of the duration of a note of the same pitch which is already sounding (or within an 8th note, whichever is shorter), it is considered a duplicate and is combined with the previous note. If it starts later than that, the first note is shortened so that it ends at the same tick at which the new one starts.

To use this command, make an Edit selection that includes the duplicate notes, then choose 'Remove Duplicate Notes' from the Event menu.

Stuck Notes

A typical problem that comes up time and time again when you are programming MIDI instruments is 'stuck notes'. Typically, this happens when you change 'patch' before the patch you are playing has received a MIDI Note Off command. If this happens, you can use the 'All MIDI Notes Off' command from the Event menu to turn off any stuck notes on connected MIDI devices.

tip ▷ The keyboard command for 'All MIDI Notes Off' is Command-Shift-Period (Mac) or Control-Shift-Period (Windows).

note ▷ The 'period' character (.) on the computer keyboard is called the 'full-stop' in UK English.

Useful Keyboard Commands

Two extremely useful keyboard commands let you apply changes to all tracks or to whichever tracks you have selected. For example, you might want to set all the track outputs or all selected track outputs to the same destination. To 'Do to All', press and hold Option-click (Mac) or Alt-click (Windows) as you make your change. To 'Do to All Selected', press and hold Option-Shift-click (Mac) or Alt-Shift-click (Windows).

When you are editing MIDI data, you will frequently need to change between Notes and Region view. There is a handy keyboard command that lets you do this: press and hold the Control key and the minus ($-$) key and this will immediately switch MIDI tracks between Notes and Regions view.

note ▷ Instrument tracks can be edited just like MIDI tracks and also let you edit automation playlists, just like Auxiliary Input tracks. Instrument tracks also support the Pro Tools 'Do to All' and 'Do to All Selected' keyboard commands.

Special Cut, Copy, Paste, and Clear Commands

The Edit menu commands, Cut Special, Copy Special, Paste Special, and Clear Special can be very useful for editing MIDI controller data. These each have three sub-menu selections, two of which let you edit either all the MIDI controller data or just the MIDI pan data. (The third sub-menu command only works with audio plug-in automation.) Conveniently, these commands work whether the data is showing in the track or not.

The Cut Special, Copy Special and Clear Special commands work identically. So the 'All Automation' sub-menu selection cuts, copies or clears all MIDI controller data and the 'Pan Automation' sub-menu selection cuts, copies, or clears only MIDI pan data.

The Paste Special sub-menu commands require further explanation. The 'Merge' sub-menu selection pastes MIDI controller data from the clipboard to the selection and merges it with any current MIDI controller data in the selection.

The 'To Current Automation Type' sub-menu selection pastes MIDI controller data from the clipboard to the selection and changes this to the current MIDI controller data type. This lets you convert from any MIDI controller data type to any other MIDI controller data type. For example, you could copy MIDI volume data and paste this to MIDI pan.

The 'Repeat to Fill Selection' sub-menu selection repeatedly pastes the MIDI controller data from the clipboard until it fills the selection – saving lots of time compared with manual pasting of selections. Simply cut or copy a MIDI region, then make an Edit selection and use this command to fill the selection. And if you have selected an area that is not an exact multiple of the copied region size it will automatically trim the last copied region that is pasted so that it fits exactly.

Marquee Zooming

Pro Tools 7 supports so-called Marquee Zooming for MIDI and Instrument tracks. This is where you drag a rectangular outline around a selection of notes using the mouse and the Edit window display zooms to encompass just this selection.

Select the Zoom Tool then Command-click (Mac) or Control-click (Windows) to make a marquee zoom selection on a single track and the display will zoom in vertically and horizontally at the same time. Take a look at the accompanying screenshot to see how this Marquee Zooming looks in action.

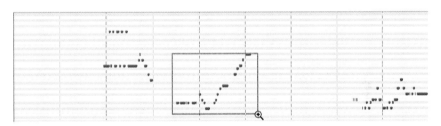

Fig. 7.8 – Marquee Zooming in action.

The group of notes encompassed by the rectangular marquee is zoomed horizontally until it fills the available space in the Edit window. The notes are also zoomed vertically if this is necessary to make them all visible within the display. Take a look at the accompanying screenshot to see how the example selection appears after using the Marquee Zooming feature.

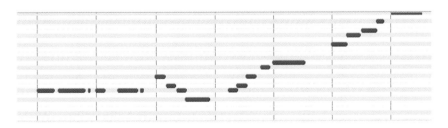

Fig. 7.9 – The Edit selection zoomed to fill the window.

note ▷ In Regions view the display zooms horizontally to encompass the Edit selection, but not vertically, so you may not see all the notes encompassed within the selection if you switch back to Notes view – depending on the vertical zoom resolution that last applied.

Keyboard Zoom Commands

Pro Tools 7 also offers a keyboard command that zooms the selection into fill the Edit window. Make an Edit selection first, then press Option-F (Mac) or Alt-F (Windows) and *all* the MIDI and Instrument tracks will zoom horizontally until the current edit selection fills the Edit window. *Selected* tracks will also zoom vertically to make sure that all the MIDI notes within the selection are visible.

Very often, you will want to zoom all the tracks out so that you can see an overview of your session. If you press Option-A (Mac) or Alt-A (Windows), this will zoom all the tracks out all the way horizontally and vertically. MIDI and Instrument tracks will automatically zoom vertically to display all notes.

Individual Vertical Zoom

Marquee Zooming and the keyboard zoom commands change the zoom levels for all the MIDI and Instrument tracks in the Session at the same time. There will be occasions when you want to have different MIDI or Instrument tracks at different vertical zoom levels.

To zoom just one track vertically, select the Zoom tool and then hold the Control key (Mac) or the Start key (Windows) as you click and drag up to zoom in vertically, or click and drag down to zoom out vertically.

Take a look at the accompanying screenshots to see how this works. The first screen shows the Zoom tool cursor near the bottom of the displayed MIDI notes. The modifier key (Control or Start) is being held down.

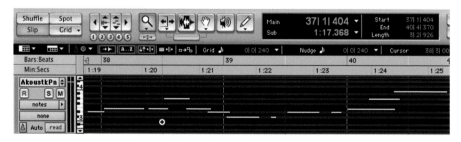

Fig. 7.10 – The Zoom tool cursor about to be dragged vertically upwards while holding the Control (Mac) or Start (Windows) modifier key.

The second screen shows the Zoom tool cursor near the top of the displayed MIDI notes. As you can see, the vertical zoom level has changed so that the display has zoomed in on the notes, making them larger.

Fig. 7.11 – The Zoom tool cursor after dragging upwards to zoom the display vertically.

note ▷   Independent vertical track zoom is not available in Regions view. This is because in MIDI Regions view, tracks scale their vertical range to always show all notes at all times, regardless of individual track vertical zoom levels or track height.

The MIDI Operations Windows

The Event menu lets you access the MIDI Operations Window, which actually includes nine different windows that you can select using the popup selector at the top of this window. These windows include Grid/Groove Quantize, Input Quantize, Restore Performance, Flatten Performance, Change Velocity, Change Duration, Transpose, Select/Split Notes, and Step Input.

To apply any of these operations you typically need to select a MIDI note or region first then choose the operation you wish to apply.

tip ▷   You can open the MIDI Operations window by pressing Option-3 (Mac) or Alt-3 (Windows) on the numeric keypad.

Grid/Groove Quantize

The Grid/Quantize window provides both standard Quantize and Groove Quantize features.

tip ▷   You can open the Grid/Groove Quantize window by pressing Option-zero (Mac) or Alt-zero (Windows).

Grid Quantize

If you record MIDI data in real time using a MIDI keyboard or other MIDI controller you will often find it helpful to correct the notes using the standard Quantize features. Quantizing the notes moves them onto the theoretically correct bar and beat locations. This can sound too mechanical, so options are provided to let you specify how much to move the notes and which notes to leave alone because they are close enough already.

You can select the note value to quantize to using the Quantize Grid popup. The default value is 16th notes. Options are provided for 'tuplets' such as triplet values (3 notes played in the time of 2) and to let you randomize the values by a percentage value that you can choose. To let you adjust the 'feel' so that notes are moved to just before or just after the theoretically correct beats, you can offset the quantize grid by a specified number of ticks.

Other options let you set a Strength value (how close to the theoretically correct values you will move the notes), a Swing value, and whether to include or exclude notes within a specified percentage of the distance from the theoretically correct values.

Fig. 7.12 – Grid Quantize.

Groove Quantize

The Quantize Grid popup also lets you choose a 'groove' to quantize to. Preset grooves include settings for Cubase Style, Logic Style, and MPC60 Style quantize 'groove' templates. If you use these to quantize your MIDI tracks, you should get similar results to those you would get using these popular sequencers.

note ▶

Sound Architect Ernest Cholakis developed the Feel Injector Templates especially for Pro Tools. These templates are designed to impart an element of human performance into your MIDI and audio sequences. Cholakis suggests that when applying Feel Injector Templates to MIDI sequences you should use a velocity sensitivity of around 75% to 80%. Apparently, the velocity sensitivity information is based on real-world dynamics, which works well for audio recordings but is a bit too wide for MIDI sequences. The timing sensitivity should be set to 100%. In each Feel Injector Template, a guide tempo is listed. This gives an indication of the original tempo on which the groove was based, so you can take this into account if you like – or you can just go ahead and apply a Feel Injector Template to a MIDI sequence running at any tempo.

Groove Quantize lets you set a randomization value to make grooves applied to multiple tracks sound more lively and realistic. You can also offset the timing, duration or velocity values – giving you a tremendous amount of flexibility to vary the grooves.

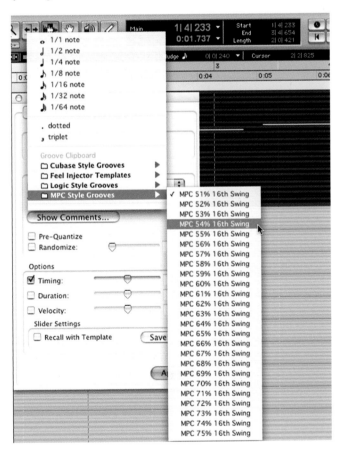

Fig. 7.13 – Using the Quantize Grid popup to select an MPC Style 'groove' value.

Input Quantize

The Input Quantize window provides all the options available in the Grid/Groove Quantize window to let you apply quantization or grooves as you are playing your MIDI data into Pro Tools from your MIDI keyboard or other controller.

Fig. 7.14 – Input Quantize window.

Restore Performance

The Restore Performance command lets you restore the original performance any time you like – even after the session has been saved or the Undo queue has been cleared. You can select which note attributes to restore – timing (quantization), duration, velocity, or pitch – and the selected MIDI notes will revert to the way they were when you originally recorded them (i.e. before any subsequent edits).

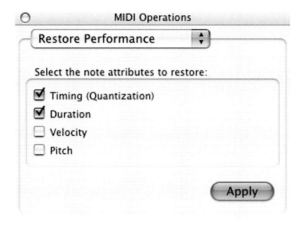

Fig. 7.15 – Restore Performance dialog.

Flatten Performance

The Flatten Performance command 'solidifies' your edits – letting you lock or 'flatten' the current state of selected MIDI notes. Flattening a MIDI Performance also creates a new 'restore to' state for specific note attributes that you can select – including duration, velocity, pitch, and timing (quantization). You might use this command, for example, if you made a series of edits that you considered to be 'correct', but you then wanted to be able to try out more edits that may not be correct – while retaining the ability to restore to this 'correct' state.

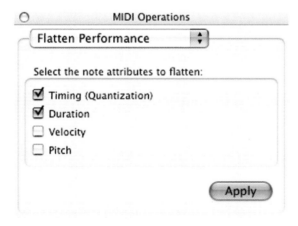

Fig. 7.16 – Flatten Performance dialog.

Change Velocity

You can change the velocities of all the notes you have selected in a variety of useful ways – adding or subtracting a specific value, scaling over the length of your selection, or even introducing a specifiable element of randomization to the values. You can use the change smoothly options to create crescendos or diminuendos.

187

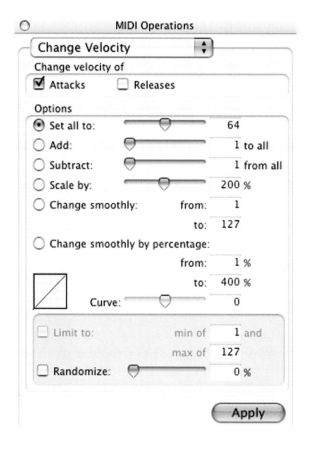

Fig. 7.17 – Change Velocity window.

Change Duration

The Change Duration window provides control over the durations of selected MIDI notes. You can use this to make melodies and phrases more staccato or more legato.

tip ▶ Press Option-P (Mac) or Alt-P (Windows) to open the Change Duration window.

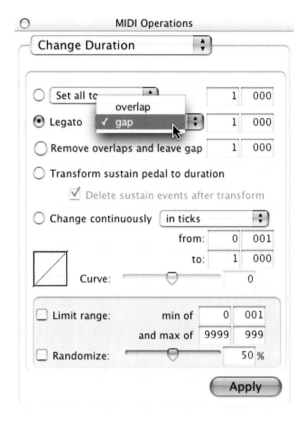

Fig. 7.18 – The Change Duration window.

To make your MIDI notes more staccato you can use the first option to set all the values to a specific short value. You can also add or subtract specific values, or scale selected values by a percentage.

To make the MIDI notes legato, use the second option and choose 'Gap' or 'Overlap' from the 'Legato' popup menu. You can then specify the length of the gap or overlap in beats and ticks.

The third option, 'Remove Overlap', removes any note overlap for all notes of the same pitch. This command is different from 'Legato' in that non-overlapping notes of the same or different pitches remain unchanged. To create a gap between previously overlapping notes, enter the length of the gap in beats and ticks.

The fourth option, 'Transform Sustain Pedal To Duration', extends the length of any notes which are sounding when the sustain pedal (MIDI controller 64) is considered to be down (with values between 0 and 63) to the point when the sustain pedal is considered to be up (with values between 64 and 127). If you select the 'Delete Sustain Pedal Events' option, this will delete any sustain pedal events within the selection.

The fifth option, 'Change Continuously', lets you change the duration of notes continuously in ticks or by percentage. You can adjust the 'Curve' slider to modify the shape of the change.

You can also limit the range of durations within your selection, and introduce a specifiable percentage of randomization to the values to help to avoid the notes sounding too 'mechanical'.

Transpose

When you have recorded a bassline, synth pad, Moog leadline, or suchlike using MIDI, you always have the option to try this using a different synthesizer sound. If you switch to a different synth 'patch' other than the one you used to record the part, this may play back an octave lower or higher than the original. Or you may decide that the part you have recorded simply sounds better in a different octave – or even a number of semitones up or down from the original. A common arranging technique is to copy the part to one or more new tracks and shift the pitch of the new tracks to form harmonies with the original. This is where the Transpose window comes to your aid, allowing you to move MIDI notes up or down by a specified number of semitones and/or octaves. You can also transpose selected notes from their original pitch to a new pitch, or move all the selected notes onto a specific pitch.

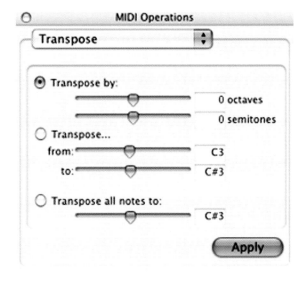

Fig. 7.19 – Transpose window.

tip ▷ Press Option-T (Mac) or Alt-T (Windows) to open the Transpose window.

Select and Split Notes

The Select/Split Notes dialog lets you select or split notes according to various rules that you can specify within the dialog.

For example, you can extract notes between a specified pair of pitches containing the notes that you are interested in. Or you can choose notes falling within a specified velocity, duration or position range. The Velocity option lets you specify a range of velocities and the Duration option lets you specify a range of durations (in beats and ticks) for selecting or splitting notes. The Position option lets you select or split all notes that fall into a range of beat and tick locations within each bar.

Once you have specified the criteria by which the notes will be selected or split, you can choose the action that will be taken when you click the Apply button. So, for example, you can choose whether selected data is copied to the clipboard or to new tracks, or whether the data is split to multiple new tracks or a single destination.

To use this feature:

Step 1. Choose the Select/Split Notes sub-menu option from the MIDI item in the Event menu.

tip ▷ Press Option-Y (Mac) or Alt-Y (Windows) to open the Select/Split Notes window.

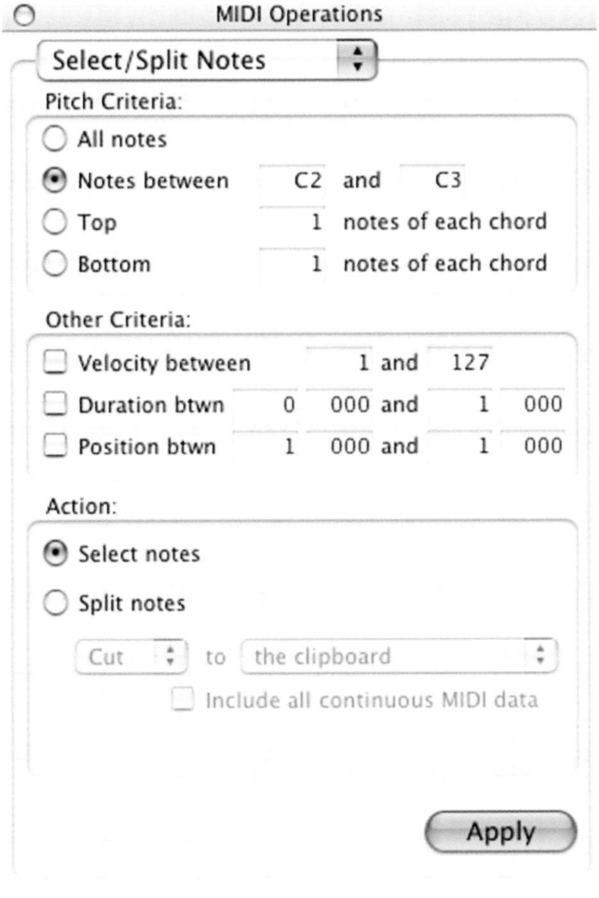

Fig. 7.20 – Split Notes dialog.

Step 2. Choose the Pitch Criteria that you want to use.

Step 3. Choose any other criteria, such as Velocity.

Step 4. Choose the Action: Select or Split.

Step 5. If you have chosen Split, a popup menu just below this lets you choose whether to Copy or Cut the data.

Step 6. The popup to the right of this lets you choose the destination that the data will be cut or copied to.

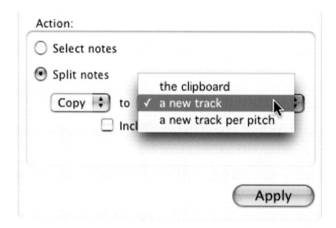

Fig. 7.21 – Split Notes Action options.

note ▷   The default is to copy the selected data to the clipboard, after which you must manually paste the data to your chosen destination.

The 'A New Track' option copies all the selected data to a single, new track. If the selection includes multiple tracks, that same number of tracks will be created. The new tracks will duplicate the original track type (Instrument or MIDI).

The 'A New Track per Pitch' option copies all selected data to multiple new tracks (one new track for each pitch). Be aware that if the selection includes multiple tracks, each track will have its data split separately.

Step 7. Enable the 'Include MIDI Controller Data' option if you want to include all the controller data associated with the split notes in the newly created tracks.

Step 8. Click Apply to carry out the selected action on the selected data.

note ▷   When you click Apply, any real-time MIDI properties associated with the track or regions are copied to new tracks created by Select/Split Notes.

Step Input

This window lets you use a MIDI keyboard or other controller to enter notes individually, one step at a time – giving you precise control over note placement, duration, and velocity. You can also use Step Input to create musical passages that are too difficult or too fast to play accurately.

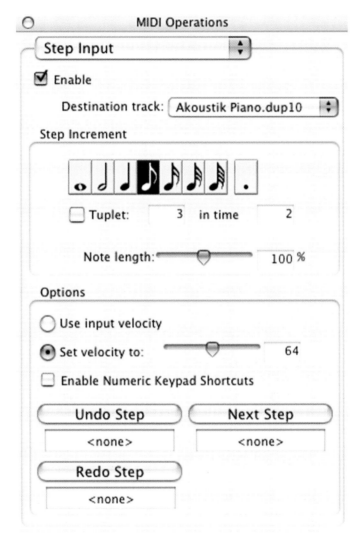

Fig. 7.22 – Step Input window.

MIDI Real-Time Properties

To compete with software such as Logic and Cubase, Pro Tools 7 now has a Real-Time Properties window that lets you adjust and apply MIDI quantization, duration, delay, velocity, and transposition parameters during playback.

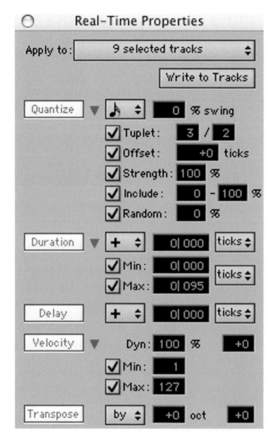

Fig. 7.23 – MIDI Real-time Properties window.

Real-Time Properties can be applied to a whole track or just to particular regions within a track. To indicate that Real-Time Properties are being applied to a track, a 'T' is displayed in the upper right-hand corner of the track's MIDI regions and in the Event List. To indicate that Real-Time Properties are being applied to a region, an 'R' is displayed in the upper right-hand corner of the region and in the Event List.

One of the advantages of using Real-Time Properties is that you can quickly try out different settings for quantization, transposition and so forth before you decide whether to make these settings permanent. When you are sure you have the settings you want, you can write these settings to the selected tracks or regions. This overwrites the corresponding MIDI properties on the selected tracks or regions and resets the Real-Time Properties display.

Real-Time Properties can be adjusted in the Real-Time Properties View in the Edit window, or in the Real-Time Properties window.

How to compensate for monitoring latency when using external MIDI devices

If you are monitoring the output of your MIDI devices through a Digi 002, Digi 002 Rack, Mbox 2, Mbox, or any of the M-Audio interfaces, your MIDI tracks will sound as though they are playing slightly later than your audio tracks. This happens because the audio coming into the audio interface from your MIDI devices has to be processed by the computer before it is routed through to your monitor speakers. This processing causes a delay, referred to as the 'latency' in the system. The latency is greater if you use larger hardware buffer sizes and can be minimized if you use lower buffer sizes. However, even at the smallest buffer size, there is still some latency. Also, reducing the buffer size limits the number of simultaneous audio tracks that you can record without errors.

note ▷ Use a larger buffer size when mixing so that you get higher track counts with more plug-ins. Use a smaller buffer size when recording audio that is monitored through your Pro Tools LE system.

tip ▷ One way to get around the latency problem is to monitor your MIDI devices using an external mixer, before it is routed to Pro Tools.

If you don't have an external mixer, another way to work around the problem is to compensate for the latency by making the MIDI and Instrument tracks play back earlier than the audio tracks so that when you monitor the audio coming from your external MIDI devices it sounds in time with the rest of your audio tracks.

You can do this using the Global MIDI Playback Offset preference. This allows MIDI and Instrument tracks to play back earlier (or later) by a specified number of samples. Using this offset only affects the way the MIDI data plays back: it doesn't alter the way that the MIDI data is displayed in the Edit window.

To configure the Global MIDI Playback Offset:

Step 1. Choose Preferences from the Setup menu and click the MIDI tab.

Step 2. Enter a negative number of samples for the Global MIDI Playback Offset value to make the MIDI tracks play back earlier than the audio tracks. Enter a positive value to make the MIDI tracks play back later.

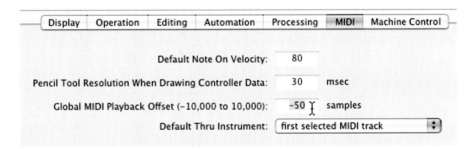

Fig. 7.24 – Entering a Global MIDI Playback Offset in the MIDI Preferences window.

note ▷ To allow for monitoring latency in Pro Tools LE, set the offset to a value that matches the H/W Buffer Size. You can check what the buffer size is by opening the Playback Engine dialog from the Setup menu.

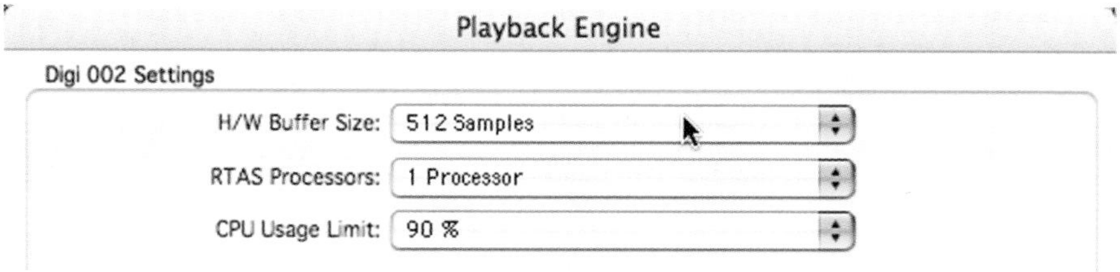

Fig. 7.25 – Playback Engine dialog showing H/W Buffer Size.

Step 3. Click 'Done' in the MIDI Preferences dialog.

note ▷

The Global MIDI Playback Offset can also be set from the MIDI Track Offsets window.

How to compensate for triggering delays when using external MIDI devices

All MIDI devices are not equal. When a particular synthesizer, such as a Yamaha model, receives MIDI data, it will take a few milliseconds before it plays the sound. Send the same data to a Roland synthesizer and this will take a different number of milliseconds to respond. Try the same thing with an Akai sampler and an EMU sampler. Almost certainly the response times before they play the audio will be different.

If the response time is very fast, say 5 milliseconds or less, and depending on the type of sound the synthesizer is playing, this may not be a problem. String sounds usually have a slow attack time, for example, so you would probably not hear anything 'wrong'. Where it does become a problem is when you are working with a MIDI device that has a very slow response time, or when you are working with sounds that have very fast attack times, such as percussion instruments.

And if you have some audio in Pro Tools that should play exactly at the same time as the audio coming from an external MIDI device, this can exacerbate the problem. Remember that it usually takes at least 5 milliseconds to trigger the notes from an external MIDI device, and it could take even longer, depending on the device. So if you have a kick drum that is being played by an audio track in Pro Tools and a kick drum that is being played by a MIDI device triggered by MIDI notes coming from Pro Tools, you would hear a 'flam' (a delay) between the two kick drums.

To compensate for delays caused by the time it takes to trigger events on a MIDI sampler or synthesizer, you can offset individual MIDI and Instrument track offsets in Pro Tools.

You can measure the triggering delay for a MIDI device assigned to a MIDI track by recording its audio output back into Pro Tools. Compare the sample locations for the recorded audio events against the original MIDI notes to calculate the delay time.

To configure a MIDI or Instrument track offset for a track:

Step 1. Choose MIDI Track Offsets from the Event menu.

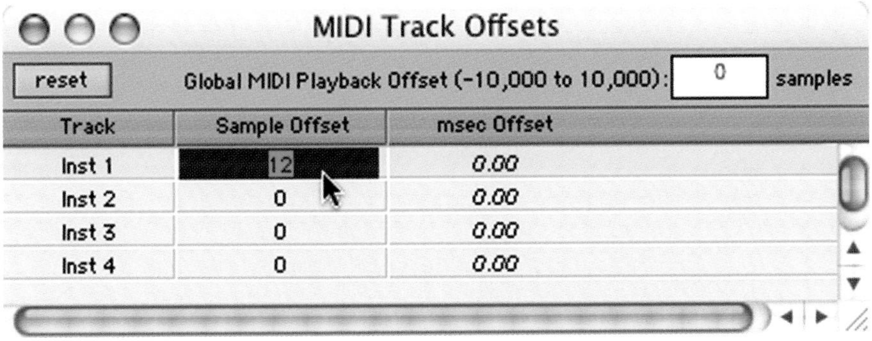

Fig. 7.26 – Entering individual MIDI/Instrument Track Offsets.

Step 2. Click in the Sample Offset column for the MIDI or Instrument track and enter the number of samples (between −10,000 and +10,000) for the offset.

Negative values cause the MIDI or Instrument tracks to play back earlier than the audio tracks; positive values cause the MIDI or Instrument tracks to play back later.

The msec Offset column shows the equivalent offset in milliseconds. This value updates when a new value is entered in the Sample Offset column, although it cannot be edited.

Step 3. Press Return (Mac) or Enter (Windows) on your computer keyboard to accept the entered offset value.

note ▷ To reset all offsets for all MIDI and Instrument tracks, click the Reset button in the upper left of the window.

tip ▷ You can also apply delays using the MIDI Real-Time Properties to offset MIDI or Instrument tracks.

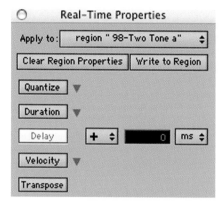

Fig. 7.27 – The MIDI Real-Time Properties can be used to delay a track.

Quick Start Editing Session

The best way to learn how to edit the MIDI, as with most things, is to record some MIDI data and start cutting, copying, pasting and using the various edit commands until you get completely familiar with how these all work. A good way to get started is to record some drumbeats and then play around with these.

Programming Beats

Probably the quickest way to get started programming beats in Pro Tools is to insert SampleTank onto an Instrument track, hook up a MIDI keyboard or percussion controller, and start tapping out rhythms on a bass drum sound, snare drum sound or whatever.

Step 1. Open a new Session.

Step 2. Get a click set up and working.

Step 3. De-select the Metronome button in the Transport window and manually set a tempo, or select the Metronome button and write tempos into the Tempo track.

Step 4. Make sure that the main display shows Bars and Beats.

Step 5. Add an Instrument track and insert SampleTank.

Step 6. Choose the drum kit you want to work with from SampleTank's library.

Step 7. Choose 'MIDI > Input Quantize. . .' from the Event menu and select an appropriate quantize grid. The default value of sixteenths works for lots of popular music – but not for everything. If you are not a musician, ask a musician how this should be set.

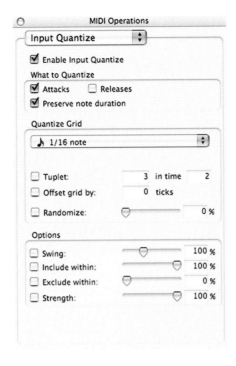

Fig. 7.28 – Input Quantize dialog.

Step 8. Record enable the Instrument track.

Step 9. Press Record and Play in the Transport window to start recording, play your MIDI controller, then press Stop when you are done.

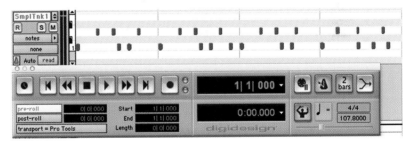

Fig. 7.29 – SampleTank Instrument track showing recorded MIDI notes.

The notes you play will all be quantized (i.e. moved) to the nearest 16th note (or whatever you have chosen in the Input Quantize dialog) to correct any notes that you have not played exactly on the right beat.

Manually Correcting Mistakes

If your playing is too inaccurate, some of the beats you play will be moved onto the wrong beat, too early or too late. You can easily drag these onto the correct Bar:Beat positions.

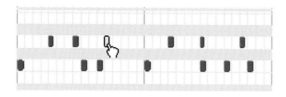

Fig. 7.30 – A MIDI note played a 16th note too early.

Step 1. Set the track to display Notes and adjust the vertical and horizontal zoom levels so that you can see the notes clearly in the display.

Step 2. Make sure that you are in Grid mode with the grid set to a suitable value – sixteenths in this example.

Step 3. Select the Grabber Tool then point, click, hold, and drag any notes that are too early or too late to their correct positions.

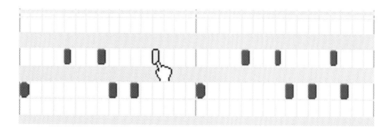

Fig. 7.31 – A MIDI note grabbed and dragged onto the correct beat.

MIDI Loop Recording

If you would like to keep playing new ideas until you find what you want, you can use Loop Record mode to record new 'takes' into a pre-defined section until you are sure that you have one or more 'good' takes.

Step 1. Before you start recording, set a Start and End point in the Transport window to define a section to loop around and select Loop Record mode from the Options menu.

Step 2. Start recording and keep playing until you feel that you have recorded at least one good 'take', then press Stop.

Fig. 7.32 – Recording in Loop Record mode.

Step 3. When you have finished recording, you can choose the best take by Command-clicking (Mac) or Control-clicking (Windows) in the playlist with the Selector Tool active.

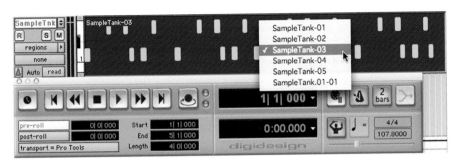

Fig. 7.33 – Choosing the best take.

MIDI Drum-machine style recording

Pro Tools can record in the way that drum machines do, where they typically loop round a 2-, 4-, 8-, or 16-bar pattern and let you keep on adding beats to the pattern.

Step 1. Set a Start and End point in the Transport window to define a section to loop around.

Step 2. Choose 'Loop Playback' from the Options menu.

Step 3. Enable the MIDI Merge button in the Transport window.

Step 4. Start recording and playing your MIDI controller.

tip ▷

> You could play the bass drum first, then add the snare on the next pass around the loop, then the hi-hat, then the cymbals, then add an extra bass drum beat into the pattern – and so on until you have what you want. And you are not restricted to drum patterns – you can play any instruments or sounds using this method.

Step 5. Press Stop when you are finished.

Fig. 7.34 – MIDI Merge recording.

tip ▷ Don't forget to choose an appropriate name for the region, such as '4-bar Drumloop'.

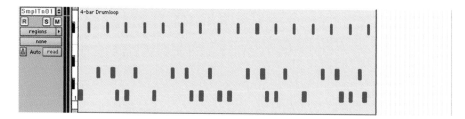

Fig. 7.35 – The 4-bar drum pattern, ready for 'looping', or, in other words, repeating.

Repeating Patterns

Step 1. Set the Grid to 1 bar again.

Step 2. Display Regions or Blocks in the track.

Step 3. Use the Grabber tool to select the 4-bar region.

Step 4. Use the 'Repeat. . .' command from the Edit menu to specify a number of repeats to create a 16- or 32-bar verse or chorus section.

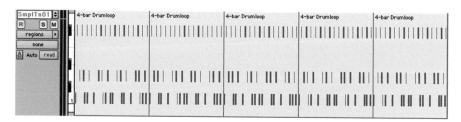

Fig. 7.36 – The 4-bar drum pattern 'looped', that is, repeated four times, to form a 16-bar section.

Mirrored Edits

If you notice a mistake in the pattern, or if you change your mind about the placing of any of the beats later on, you can use Mirrored MIDI Editing to change all instances of the region to be correct.

For example, you may decide that the last bass drum beat in the four-bar pattern should be placed a 16th note earlier in the bar.

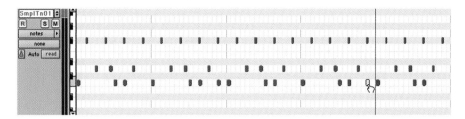

Fig. 7.37 – The last bass drum note in the 4-bar pattern is highlighted in this screenshot.

Step 1. Change the display to show the MIDI notes and make sure that the display is sized and zoomed to allow you to see the notes clearly.

Step 2. Select 'Mirror MIDI Editing' from the Options menu.

Step 3. Make sure you are in Grid mode and that the grid is set to a suitable value – in this case 16th notes.

Step 4. Select the Grabber tool, point, click, hold, and drag the note to a new position.

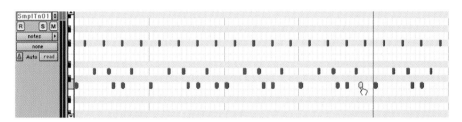

Fig. 7.38 – The last bass drum note in the 4-bar pattern has been moved a 16th note earlier in the bar.

This edit is automatically reflected throughout all the repeated 4-bar drum patterns based on this pattern, saving you the trouble of editing the last bass drum in every pattern.

You can make a whole variety of edits like this – dragging notes to different pitches, lengthening or shortening notes using the Trimmer tool, and so forth. You can also apply the various MIDI operations such as Quantize, Transpose, Change Velocity, and Change Duration to selected notes or groups of notes.

note ▷ | If you select groups of MIDI notes by drawing a rectangle around them using any of the Grabber tools, only the note data is selected. On the other hand, if you select a bunch of MIDI notes using the Selector tool, all the controller data in the track will also be selected.

What you have learned

MIDI note locations and lengths are expressed in terms of bars, beats and ticks.

MIDI data can be edited graphically or using the Event list.

Useful editing commands let you remove duplicate notes, stop stuck notes from playing, and cut, copy, paste, and clear MIDI data.

To present the data in the Edit window more appropriately, there are various ways that you can zoom the display.

To quantize data, change velocities or durations, transpose notes, select and split notes, you can use the various MIDI Operations windows.

A Step Input window lets you enter MIDI data using the computer keyboard.

You can apply quantize, velocity, duration, transpose delay properties to tracks or regions in real time (i.e. on playback without changing the original data) using the MIDI Real-Time Properties features.

You can compensate for monitoring latency and/or triggering delays when using external MIDI devices.

If you have followed the Quick Start Editing Session, you should be able to program basic drumbeats or other musical parts, loop the patterns and manually correct mistakes.

Recording Audio

Pro Tools is an excellent system to use for recording audio. With LE and M-Powered systems you can record and play back up to 32 tracks of audio using qualified hard drives (check the compatibility section on the Digidesign website to see which drives Digidesign has tested). The big advantage that digital audio recording has over analogue systems is that making copies once you are in the digital domain does not degrade the audio, building up layers of noise and distortion on each subsequent copy – as is the case with analogue tape systems.

One of the keys to making good recordings is maintaining a good signal-to-noise ratio. So record using levels as 'hot' as you can get them without overloading any of the components in the recording chain. If you are recording 16-bit, this is very important, but with 24-bit systems it is less crucial to keep the signal levels at their maximums. 24-bit systems can record a much wider range of signal levels, so even if the signal is relatively low, it may still be using, say, 16 of the 24 available bits, and if this audio is intended to sound quiet in the mix, you will not hear any low-level noise.

Preparing to Record

There are several preparatory steps that you may need to take before you start to record audio, so let's look at these first.

Monitoring Levels vs Input Gain Levels

One of the first things you need to be clear about is that the volume and pan controls in Pro Tools only affect the monitoring (listening) levels – they do not have any effect on the input levels that you are recording.

The place to adjust input levels is at the audio interface. Use the controls on your interface's microphone, instrument or line level inputs to set the input levels high enough to make full use of the dynamic range available in Pro Tools while avoiding clipping of the input waveform, which causes distortion that you cannot remove later.

The way to adjust input gain levels is to sing or play the loudest sound that you wish to record and turn the input level controls up until the meters indicate clipping. Then back off the input gain control until this clipping disappears. Then back it off a little more in anticipation of the fact that many musicians actually play or sing even louder when recording rather than rehearsing – in their enthusiasm for the 'take'.

tip ▷   Getting the input levels right is one of the main 'keys' to making quality recordings

Auto Input vs Input Only Monitoring

You also need to be aware of the two different ways in which the input signals can be monitored while playing back, while recording, or with the transport stopped, namely the Auto Input and Input Only monitoring modes.

Let's look at the default Auto Input mode first. When playback is stopped, the track monitors whatever is coming into its audio input. Now let's consider what happens when you are overdubbing and reach a punch-in point. In this case, you will normally want to hear whatever has been recorded up to the punch-in point. Then you want to hear the new audio that is coming in. When you reach the punch-out point you want the monitoring to switch back so that you can hear the existing track material again. So this is exactly what Auto Input mode does.

Sometimes you don't want to hear the original material that was recorded, you want to hear the incoming audio at all times. This is what Input Only Monitoring does. You will find a menu command for this in the Track menu. To warn that you have switched to this mode, the Input Monitor Enabled Status indicator in the Transport window lights green when Input Only mode is enabled.

tip ▷   To switch between Auto Input and Input Only monitoring, use Option-K on the Mac or Alt-K on Windows.

note ▷   When you record-enable audio tracks, their volume faders turn bright red. This is to remind you that different fader levels can apply during recording and playback if you de-select 'Link Record And Play Faders' in the Operation page of the Preferences window. With this preference de-selected, if you adjust a fader when a track is record-enabled and then turn off record enable for the track, the fader returns to its playback level.

   When the Operation preference for 'Link Record and Play Faders' is selected, Pro Tools does not keep track of different record and play levels for audio tracks. In this case, record enabling an audio track has no effect on the fader level for the track. This lets you maintain a consistent mix regardless of whether you're recording or just listening.

Latency

Because the audio coming into Pro Tools LE has to be processed by the computer on its way through the system, before it is routed to your monitoring system, you will hear a delayed version of the input through your monitor speakers.

This can be very off putting for a musician who is trying to overdub to an existing performance, because the music already recorded plays back at the correct time while the new performance is delayed with respect to this.

It is possible to reduce the size of the hardware buffer to minimize this so-called 'latency' delay. You can do this in the Playback Engine dialog that you can access from the Setup menu. The disadvantages here are that at the smallest buffer sizes there is still some latency, and reducing the buffer size limits the number of simultaneous audio tracks that you can record without encountering performance errors.

Using an external mixer

You can avoid the problem altogether by monitoring the recording source with an external mixer, before it is routed to Pro Tools. Plug your microphones or instruments into the mixer and you will hear the sound directly through a connected monitor system – with no audible delay. At the same time, route the audio to your Pro Tools interface so that you can record it.

tip ▷ If you are monitoring through an external mixer while recording, don't forget to defeat the monitoring (pull the fader down or engage the mute button) for each track that you are recording in Pro Tools. Otherwise you will hear the delayed version of the audio mixed in with the un-delayed version.

Zero Latency Monitoring

If you are using an Mbox or Mbox 2, you can use the Mix control on the front panel to set up the output mix that is sent to the monitor system. The Mix control lets you balance between the audio coming into the interface's analogue input and any audio coming back from Pro Tools. Because the audio coming into the interface's analogue input is passed directly to the Mbox's monitor outputs, without going through the computer, there is no latency. Again, you should defeat the track monitoring in Pro Tools (pull the fader down or engage the mute button) to prevent audio being passed through Pro Tools and back out to the monitors where it would interfere with the directly monitored audio.

Low Latency Monitoring

If you are using a Digi 002 or Digi 002 Rack, these systems have a special Low Latency Monitoring option that can be used to record with an extremely small amount of monitoring latency. When you select this from the Options menu, all tracks with inputs set to an audio interface (as opposed to a bus), and outputs set to Output 1 or Output 2, will use Low Latency Monitoring.

note ▷ When Low Latency Monitoring is enabled, any plug-ins and sends assigned to record-enabled tracks (routed to Outputs 1–2) are automatically bypassed.

Record Enabling

Each track in Pro Tools has a Record Enable button that must be selected (lit up red) before you put Pro Tools into record mode using the Transport controls. You can record simultaneously to multiple tracks by record enabling your choice of audio, Instrument or MIDI tracks. When you have clicked the record-enable button for the first track you want to set up, you can add more choices by Shift-clicking on more track record-enable buttons as you figure out which you want to use.

tip ▷ If you enable the Latch Record Enable Buttons preference that you will find in the Operations section of the Preferences window then you don't need to Shift-click the Record Enable buttons to enable additional tracks – just click each in turn that you want to be enabled. They will all stay enabled until de-selected. And if you want to record-enable all the tracks in your session, if you are recording a band, for example, then Alt-clicking (Windows) or Option-clicking (Macintosh) will do this.

Record Safe Mode

If you are worrying that you might accidentally put the wrong track or tracks into record and mess up a previous recording, then be assured that it is very unlikely that this could happen with Pro Tools. Any new recordings are made to new files by default, so the previous files will still be there in your Regions list and in your session's audio folder. This is what is referred to as 'non-destructive' recording. Nevertheless, Pro Tools does have a 'destructive' recording mode in which new recordings replace previous recordings into the same files.

So, to prevent accidents, Pro Tools provides a Record Safe mode on a per track basis that prevents tracks from being record-enabled. Simply Control-click (Windows) or Command-click (Macintosh) the track's Record Enable button and this will become greyed out and won't let you enable the track to record. If you change your mind, just do this again to get out of Record Safe mode.

tip ▷ Hold the Alt (Windows) or Option key (Macintosh) at the same time and all tracks will be affected. And if you hold the Shift key as well, just the currently selected tracks will be affected.

Other Record Modes

Pro Tools actually has four different record modes. The default record mode is non-destructive, as mentioned above. In the Options menu you will find three other record modes: Destructive Record, Loop Record, and QuickPunch. You can switch between these by Right-clicking (Windows and Macintosh) or Control-clicking (Macintosh) the Record Enable button in the Transport window. When you cycle through these modes, the Record Enable button changes to indicate the currently selected mode, adding a 'D' to indicate Destructive, a loop symbol to indicate Loop Record, and a 'P' to indicate QuickPunch.

Destructive mode works like a conventional tape recorder where new recordings onto a particular track replace any previous recordings to that track. There is little justification for using this mode unless you are running out of hard disk space to record to.

Loop Record mode lets you record multiple takes into the same track over a selected time range. Each successive take will appear in the Region List and can be placed in the track using the Takes List popup menu. This is very useful when a singer or musician is trying to perfect a difficult section.

QuickPunch lets you manually punch in and out of record on record-enabled audio tracks during playback by clicking the Record button in the Transport window. This is useful when the engineer/producer wants to decide 'on-the-fly' which bit of a performance to replace while the singer or musician plays throughout the session.

Recording through plug-ins

It is always very tempting to want to record into Pro Tools while adding a plug-in effect. Unfortunately, there are various reasons why this is not always a good thing to do.

For a start, the additional processing via the plug-in adds a further latency delay that can be considerable with certain plug-ins.

Also, real-time plug-in effects are applied to the incoming signals before the record monitoring fader, but after they have been recorded onto your hard disk. The way this works is that when you insert a plug-in onto an audio track then record audio to this track, you will hear the sound of the audio being processed by the plug-in through your monitors – but this processed audio will not be recorded to disk.

To record the processed audio to disk you have to:

1. Create an Auxiliary Input.
2. Connect your audio source to this Auxiliary Input.
3. Insert the plug-in effect onto this Auxiliary Input track.
4. Route the output from this Auxiliary Input to the audio track to which you want to record.

note ▶ For these reasons, it is more usual to apply plug-in effects to your recordings after you have recorded the basic tracks. You can either apply AudioSuite plug-ins or bounce audio tracks containing real-time plug-ins to disk to create new files, or simply apply real-time plug-ins to create effects in real time within your mix.

Your first Audio Recording

OK, let's get started recording some audio.

Step 1. Launch Pro Tools and choose New Session from the File menu.

Step 2. In the New Session dialog, choose the 44.1 kHz sample rate, change the bit depth from the default 24-bit to 16-bit, leave the default Broadcast WAVE (BWF) file format selected, name the session, choose where to save it, then click 'Save'.

Fig. 8.1 – New Session dialog.

Step 3. Select New… from the Track menu and create 1 mono Aux input for the click, 1 mono Audio track to record onto, and 1 stereo Master Fader.

Step 4. Insert the click plug-in onto the Aux track, rename this as the Click track, and make any adjustments to the sound and level of the click.

Step 5. Open the I/O Setup dialog from the Setup menu and name the first pair of inputs to correspond to the names of your microphones, such as SM57 and SM58 as in this example.

(If you only have 1 microphone, name the first of these. If you have more, name as many as you like. If you don't have any microphones then connect a musical instrument such as an electric guitar or a MIDI device such as a sampler or drum machine instead.)

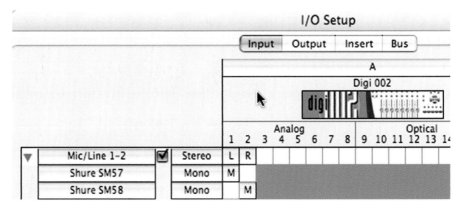

Fig. 8.2 – Naming the inputs in the I/O Setup window.

Step 6. For the Audio track, choose the track input using the popup track input selector, then record-enable the audio track.

Step 7. Sing or speak into your microphone (or play your instrument) at the loudest level you expect to use while you adjust the input level to your Pro Tools interface, keeping this as high as possible while avoiding clipping.

Step 8. Open the Transport window and use the popup selector to set the display to Bars:Beats then open the Big Counter from the Window menu so that you can see the bars and beats more clearly when you run the sequence.

tip ▷

> At this stage your screen should look similar to the screenshot below. If not, take a moment or two to arrange everything neatly on screen – you will always find a neatly arranged set of windows much faster to work with than an untidy set of windows.

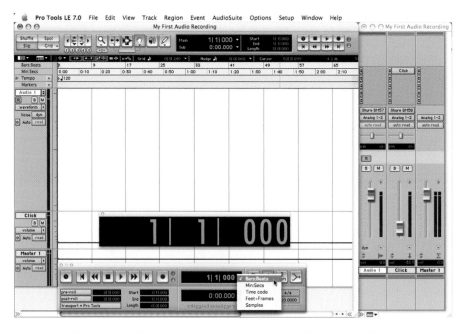

Fig. 8.3 – Pro Tools LE ready to record a mono audio track while listening to a click.

Step 9. Make sure that Destructive Record, Loop Record, and QuickPunch are all de-selected in the Options menu and that the Click in the Transport window is enabled.

Step 10. Press the Return key on your computer keyboard to make sure you are at the beginning of the session and that no start and end times are selected. You may find it helpful to set up a couple of countoff bars so that you are cued to start recording at Bar 1.

Step 11. Click Record in the Transport window to enter Record Ready mode then click Play to begin recording.

Count out loud each beat as it flashes up in the Big Counter, just like this, saying:

'1, 2, 3, 4;'

'2, 2, 3, 4;'

'3, 2, 3, 4;'

'4, 2, 3, 4.'

note ▶ This is exactly how musicians count the beats in the bars, so non-musicians who want to use Pro Tools efficiently and to work with musicians effectively should find that counting in time with the click like this helps their understanding of bars and beats and what a musician or singer has to do to keep in time with a click. For musicians, of course, this should be a trivial exercise.

tip ▷ If the default tempo of 120 BPM is too fast for you to say these numbers comfortably, just slow the tempo down a little before you start recording.

If you already started recording and you know you messed up, just press Command-Period (.) (Macintosh) or Control-Period (.) (Windows) while Pro Tools is still running. This removes the audio recorded up to that point from your hard drive and deletes the region from the track's playlist in the Edit window.

If you stopped Pro Tools running when you realized your mistake you can use the Undo command from the Edit menu to discard the take.

Step 12. Click Stop in the Transport window.

That's it! Your first audio recording has been written to disk and you will see it as an audio region in the Edit window and in the Region List.

First Fixes

Maybe you got it exactly right first time – and maybe you didn't. Listen carefully and be honest. Did you make a slip here and there and speak a little early or a little late? If you did you will be in very good company. Even the best and most talented performers make mistakes some of the time – just less often than the rest of us. So we had better talk about how to fix mistakes during the recording process so that you have something recorded which is worth editing, mixing, and mastering later on.

Recording over an existing region

Now let's suppose that you recorded a generally good 'take', but messed up Bar 2. All the numbers should correspond exactly to metronome clicks. You can tell that you messed up if any of the numbers that you speak falls in between two clicks of the metronome. You can see this better if you change the grid to quarter notes.

Here's how to fix this:

Step 1. Switch to Grid Mode and set the grid value to 1 bar.

Step 2. Choose the Selector Tool, then click somewhere near Bar 2 in the Edit window to insert the cursor into the waveform display. The blinking cursor (the vertical line that blinks on and off) will be positioned exactly at the beginning of Bar 2.

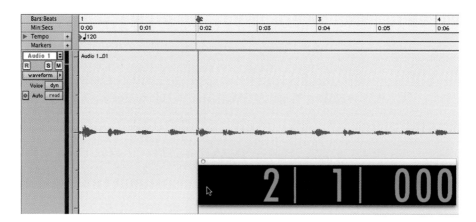

Fig. 8.4 – The cursor positioned in the playlist's waveform display in Audio track 1.

Step 3. Make sure that the click and the countoff bars are still enabled and that the track is record-enabled.

Step 4. Press Record then Play in the Transport window, pick up your microphone, wait for the countoff to complete then say '2, 2, 3, 4' into the microphone again. When you reach the end of Bar 2, just press the spacebar to stop. And with a bit of luck, you'll get it right this time! Your Edit window should look something like the screenshot below if you set the track to Jumbo size as I have done.

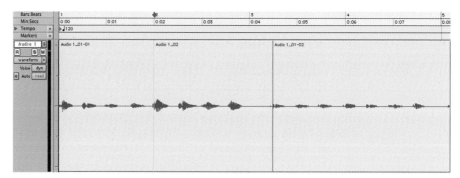

Fig. 8.5 – Recording over Bar 2.

Notice that I didn't stop exactly at the beginning of Bar 3, I over ran a little, covering up the beginning of the next beat with a silent piece of audio.

First Edits

So now it's time for your first edits.

Step 1. Choose the Trim tool.

Step 2. Make sure you are in Grid mode then trim the end of the region you just recorded so that it finishes at the end of Bar 2/beginning of Bar 3.

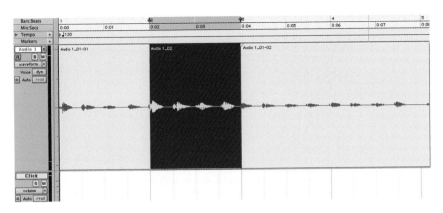

Fig. 8.6 – Trimmed region.

Step 3. Now trim the original region to end exactly at the beginning of Bar 5.

Step 4. Change to the Grabber tool, select the first region and shift-click to select the other two regions.

Step 5. Use the Group command from the Region menu to group these as one so the region can easily be moved around along the timeline. Notice the small icon that is added to the bottom left corner to indicate that this is a grouped region that can be ungrouped.

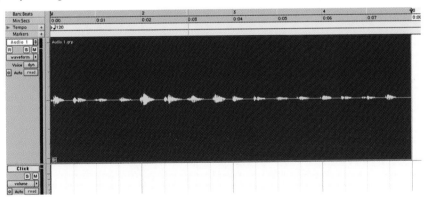

Fig. 8.7 – Grouped regions.

Separating and Moving Regions

When you listen carefully, you realize that the positioning of the words could be a little more accurate. The fix for this is to separate each word into its own region, then move these individually until they line up more accurately with the bar lines.

Step 1. Make sure the grid is set to quarter notes and select the region containing the audio.

Step 2. Choose 'Separate Region On Grid' from the Edit menu. In this example, it makes sense to separate the regions according to the grid lines, because we want each quarter note (containing one word) to be separate.

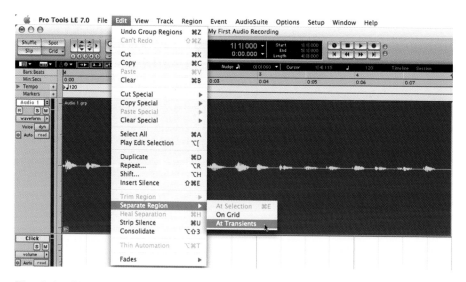

Fig. 8.8 – Using the Separate Region commands.

Step 3. Zoom in so that it is easier to see the beginning of each bar line.

Step 4. Play from the beginning of the region and listen out for the first region that is off the beat, then identify this visually, and select it.

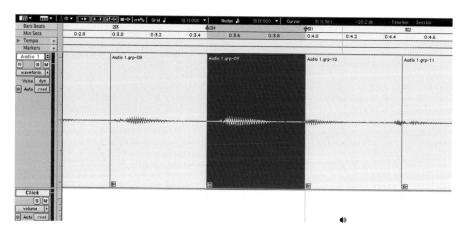

Fig. 8.9 – Beat 4 of Bar 2 sounds and looks 'off the beat'.

Step 5. Increase the vertical zoom level so that you can see the start of the audio more clearly.

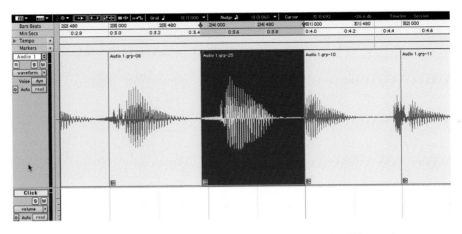

Fig. 8.10 – After increasing the vertical zoom then using the Trim tool to move the start of the region to the left, the start of the word is revealed a little way before the beat – yet the word sounds late.

Step 6. Change to Slip mode and use the Trim tool to make the region encompass all the audio that 'belongs' within that region – that is, the whole of the word.

note ▷

In this example, when I increased the vertical zoom and looked for the start of the word, I could see that this actually started earlier than the beginning of the beat – despite the fact that the word 'four' sounded late (I had held the 'fff' sound at the beginning of the word a little too long). So I actually had to use the trim tool to reveal the audio before the start of beat 4 then move the whole of this region to start a little earlier than where it was recorded to get it to sound right. This will happen to you at some time or other if you are editing audio performances, so watch out for it by zooming in and listening carefully.

Step 7. Use the Grabber tool in Slip mode to move the region along the time-line until it sounds in time with the click. In this example, dragging the region a little to the left cured the problem.

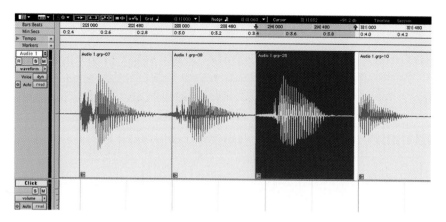

Fig. 8.11 – Moving the late-sounding word a little way to the left fixes the problem.

Now some of you may be wondering why I didn't use the command to 'Separate Region At Transients'. The transients are the peaks at the beginnings of the sounds so in theory if you separate the regions each time a new transient appears, that would put all of the separations in the correct places, immediately before the transient at the start of each spoken word. I did try this, but what happened was that the audio was often split two or three times within each word. This happens because there may be a very quiet transition between vowels and consonants that is interpreted as silence – or there may even be a small gap. Of course you can take two or more adjacent regions and use the Heal Separation command to make them join up into one region again – but this takes time. So I decided that the best choice here would be to separate on the grid then trim these as explained and move them individually into position as necessary. Another point to note from this example is that the best way to make edits is to zoom in far enough vertically as well as horizontally so that you can see what is happening with the waveform and to also use your ears to listen carefully to what is going on – adjusting the volume of your monitoring system until it is loud enough to hear the quietest sounds.

Recording multiple 'takes'

Pro Tools offers yet another method that you will find very useful when trying to perfect a recording.

Each time you make a recording, this is called a 'take'. That's the jargon that has been used in studios for decades. For example, you might record a take to Audio track 1, then record subsequent takes into tracks 2, 3, 4, 5 and so on while trying to record that perfect guitar solo or vocal performance. Of course you would have to spend a little time creating a new track and setting it up ready to record in between takes if you used this method – making it less spontaneous and running the risk of the performer 'going off the boil'.

As we have seen, you can always keep recording new takes into the same track – replacing part or all of the audio in that track. And the previous audio regions will still be available in the Regions List from which you can either delete them or drag them back into audio tracks to use later on.

But Pro Tools provides yet another way to non-destructively record a new take into the same track – by creating a new playlist for that track first.

Step 1. Click and hold the Playlist Selector popup just to the (top) left of the track meter in the Edit window.

Fig. 8.12 – Click and hold to reveal the popup Playlist Selector.

Step 2. Select 'New…' then let go of the mouse.

Fig. 8.13 – Creating a new playlist.

Step 3. Name the playlist in the dialog box that appears.

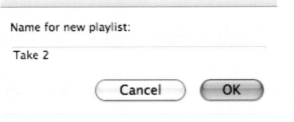

Name for new playlist:

Take 2

Cancel   OK

Fig. 8.14 – Naming the new playlist.

A new, empty, playlist appears in the track in the Edit window.

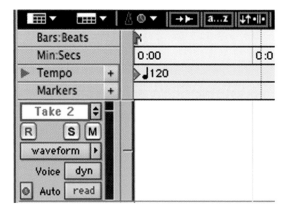

Fig. 8.15 – Playlist empty
and ready for you to
record into again.

Step 4. Record your next take into the track using this second playlist this
time.

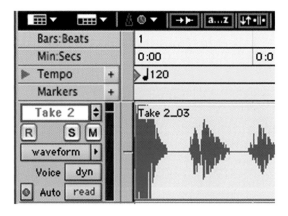

Fig. 8.16 – Take 2
recorded into the new
playlist.

tip ▷ You can always bring the first (or any previous) playlist back into the track
by selecting this using the Playlist Selector again.

Fig. 8.17 – Selecting the first
playlist again.

Step 5. You can create as many of these playlists as you are likely to need and use these to record as many takes as you have space for on your hard drives.

note ▷ When you select a playlist, the regions that previously appeared in the track are recalled. All the regions from all the playlists are always available in the Region List, from where they can be dropped into any playlist on any track.

Punch Recording

Earlier, we saw how easy it is to place the cursor at any Bar:Beat location in the waveform display and record over an existing take. And you don't have to be too careful about where you start recording from as long as it is before the audio you want to replace because you can trim the newly recorded audio afterwards so that it just replaces the section you are interested in.

However, anyone who is more familiar with using a multi-track tape recorder may feel more comfortable with the idea of setting up specific punch-in and punch-out points so that Pro Tools only records between these exactly specified locations.

Step 1. Make sure that the Edit and Timeline selections are linked. (Select Link Timeline and Edit Selection from the Options menu or make sure that the Link Timeline and Edit Selection button above the rulers is highlighted.)

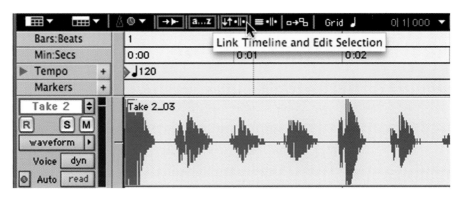

Fig. 8.18 – Engaging the Link Timeline and Edit Selection button.

Step 2. Choose the Selector Tool and drag the cursor over the range of audio in the playlist that you want to replace.

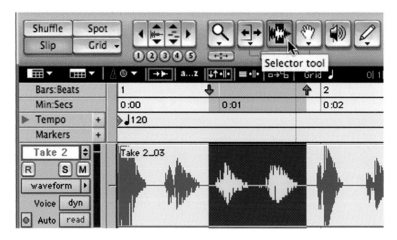

Fig. 8.19 – Use the Selector Tool to select a range of audio in the playlist.

tip ▷ If you prefer, you can type the exact locations of the punch-in and punch-out points as Start and End times in the Transport window.

Step 3. Record-enable the track or tracks onto which you wish to record.

tip ▷ You may also find it helpful to set a specific pre-roll time in the Transport window. If a pre-roll time is set, playback begins at the pre-roll time and proceeds to the start time (the punch-in point), where recording begins. When the end time (the punch-out point) is reached, Pro Tools automatically switches out of Record mode.

Step 4. When you are ready to begin recording, click Record then Play in the Transport window.

note ▷ If you are recording in the default non-destructive mode, a new audio file is written to your hard drive and a new audio region appears in the record track and Region List. If you are recording in Destructive Record mode, the new audio overwrites the previous material in the existing audio file and region.

Step 5. When you have finished recording, click Stop in the Transport window.

Loop Recording

Pro Tools also provides a Loop Record mode that lets you keep recording over the same section while creating multiple takes.

Step 1. Make sure that Loop Record is selected from the Options menu so that the loop symbol is showing in the Transport window's Record Enable button.

Step 2. If you are going to set a record or play range by selecting within a playlist, make sure that the Edit and Timeline selections are linked. (Select Link Timeline and Edit Selection from the Options menu or make sure that the Link Timeline and Edit Selection button above the rulers is highlighted.)

Step 3. Choose the Selector Tool and drag the cursor across the range of audio in the playlist that you want to loop record over. Alternatively, you can set Start and End times for the loop in the Transport window.

note ▷ The Loop Record selection must be at least 1 second long.

Step 4. Make sure that the track or tracks you want to record onto are record-enabled.

tip ▷ Although you can set a pre-roll time that will be used on the first pass and a post-roll time that will be used on the last pass, I recommend that you select a loop range that includes some time before and some time after the range you wish to record over. Later, you can trim back the recorded takes to the proper length with the Trim tool.

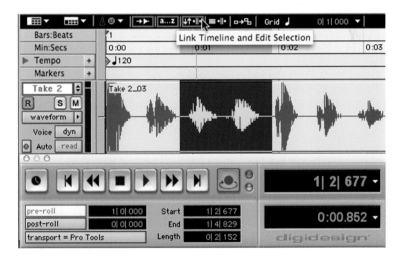

Fig. 8.20 – Getting ready to Loop Record.

Step 5. When you are ready to begin recording, click Record then Play in the Transport window.

Step 6. When you have finished recording, click Stop in the Transport window.

All the takes are recorded into a single audio file with sequentially numbered regions defined for each take. The most recently recorded take is left in the record track. All of these takes appear as regions in the Region List and you can audition takes on their own, from the Region List, or from the Takes List popup menu.

Selecting Takes

To select a different take from the Region List:

Step 1. Use the Time Grabber tool to select the current take in the Edit window.

Step 2. Command-drag (Macintosh) or Control-drag (Windows) another take from the Region List into the playlist. This replaces the previous take and snaps exactly to the correct location.

note ▷ Each region created using punch or loop record is given an identical start time (the User Time Stamp). This feature makes it possible to select and audition takes from the Takes List popup menu. You can even select different takes while the session plays or loops.

To access the Takes List popup menu so that you can select a different take:

Step 1. Make sure that the Selector tool is engaged.

Step 2. Select the take currently residing in the track, then Command-click (Mac) or Control-click (Windows) anywhere in the region. The popup menu that appears contains a list of regions that share the same User Time Stamp.

Step 3. Choose a region from the Takes List popup menu. This replaces the previous take and snaps exactly to the correct location.

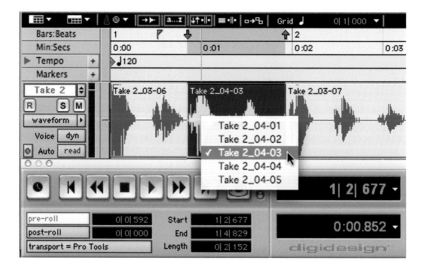

Fig. 8.21 – Choosing a take from the Takes List popup.

QuickPunch

QuickPunch lets you manually punch in and out of record on record-enabled audio tracks during playback by clicking the Record button in the Transport window.

note ▷ QuickPunch limits the number of tracks and plug-ins that you can use at the same time. The reduced track count when using QuickPunch in Pro Tools LE occurs because it requires two voices per QuickPunch track. (Of course, CPU processing power will also be a factor in how many QuickPunch tracks you can finally get.) In practice, you can record up to 16 audio tracks with QuickPunch if you are already using the maximum number of mono tracks supported by your specific Pro Tools LE system. If you need to record more than 16 tracks using QuickPunch, you will have to reduce the number of other tracks to allow the CPU to handle the extra load.

To use QuickPunch mode:

Step 1. Select QuickPunch mode from the Options menu. The letter 'P' will appear in the Record Enable button in the Transport window.

Step 2. Make sure that the track (or tracks) that you want to record onto are record-enabled.

Step 3. Press the spacebar or click the Play button to start playback.

Step 4. When you reach the location at which you want to start recording, simply click the Record Enable button in the Transport window. Pro Tools will immediately switch to monitoring the input (if it is not already in Input Monitoring mode) and will start recording.

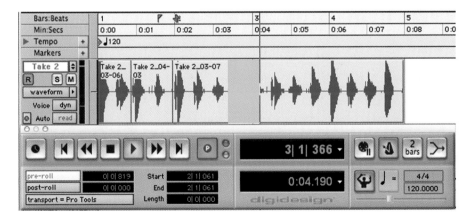

Fig. 8.22 – Recording in QuickPunch mode.

Step 5. When you are finished recording, just click Stop in the Transport window or press the spacebar to stop recording and playback.

Or,

Click the Record Enable button a second time to stop recording while continuing to play back. Then, if you want to record at other places along the timeline, all you need to do is to click the Record Enable button on and off as you please.

note ▷ QuickPunch instantly switches from monitoring the input to monitoring the playback when you punch-out of recording (which is not the case with the other record modes).

Saving your Sessions

As you build a session by recording tracks and making edits and changes to controls, you should save your work regularly. Get used to using the keyboard commands to do this (Command-s on the Mac, Control-s in Windows) and get into the habit of saving every time you have recorded a new track or made any significant edits or changes to the controls.

note ▷ The File menu offers three ways to save sessions:

1. 'Save Session' saves the currently open session file, leaving it open for you to continue working.
2. 'Save Session As' creates a duplicate session file with the name you choose and leaves the duplicate open for you to continue working. It does not create a new Audio Files or Fade Files folder. This can be useful if you want to experiment with different arrangements in the session without affecting the original session.
3. 'Save Session Copy In' saves copies of the session file and the files that you are using in the current session. This can be useful for creating a final copy of the session that does not include audio files or fades you are no longer using.

Tidying up your Sessions

When you have finished recording for the day you should spend a few minutes tidying up your session before backing it up.

Step 1. Open the Regions List and choose 'Select Unused Audio Except Whole Files' from the popup menu at the top of the Regions List.

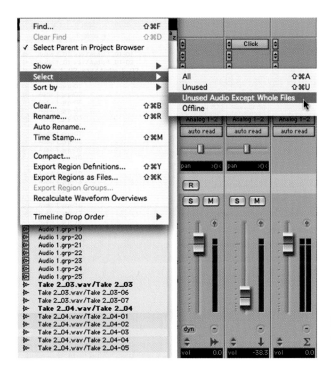

Fig. 8.23 – Selecting Unused Audio Regions.

Step 2. Take a look at what you have selected to make sure that there are no regions selected that you want to keep for any reason. For example, you may want to keep some or all of the takes that are not currently residing in active playlists.

Step 3. Choose 'Clear...' from the Regions List popup menu and click 'Remove' in the dialog that appears.

Step 4. Check that all the audio files and regions have suitable names. If not, select each file or region and choose the 'Rename...' command from the Regions List popup menu to bring up the Name dialog and type a suitably descriptive name.

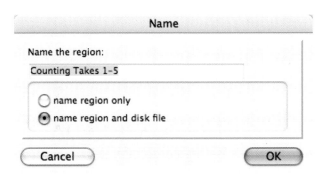

Fig. 8.24 – The Name dialog.

Step 5. Now save your Session and make a backup.

227

What you have learned

If you have followed this chapter carefully and thoroughly you should be able to record anything into Pro Tools. Once you can record successfully onto one mono track, you should have no problem recording onto a stereo track, or onto multiple mono or stereo tracks in whatever combination.

The key points are that you need to adjust the input levels on your microphone preamplifiers, instrument-level or line-level inputs to avoid clipping and to maximize the signal-to-noise ratio; be aware of the monitoring issues, how to handle latency delays, how to set listening levels differently for recording and playback; and know how to choose the most suitable recording mode.

One of the main things that you should be getting used to is that you can keep recording over existing recordings (or, more correctly, over the regions that represent these recordings) without destroying the earlier recordings – unless you choose to use Destructive mode.

You should appreciate that you need to decide on a strategy for recording multiple takes: you can either create new tracks for new takes or you can create new playlists, which is faster and simpler.

You should be reasonably familiar with how to handle punch-ins and loop recording.

You should also understand that it is essential to save your session regularly, take time out at the end of the session to tidy this up, then make a backup copy.

Editing Audio

Overview

One of the best things about Pro Tools is its incredibly flexible and accurate audio editing capabilities. Digital audio editing has allowed audio editors to develop their techniques significantly since the days of magnetic tape editing. The fact that edits can be made without affecting the original audio permanently (so-called 'non-destructive' editing) allows editors to be fearless when it comes to trying out creative ideas. The speed with which edits can be made and the accuracy – down to sample level in the case of Pro Tools – has also raised the stakes creatively.

'Destructive' editing changes the original audio – which is fine if this is what you really want to do. But it makes lots more sense to use non-destructive editing to make your edits with no fear of losing original audio – and you can always create a new audio file containing the edited audio if you want to gather it all together in one file for any reason (maybe to export it later). Destructive editing only really makes sense on a hard disk system if you are running short of space on your hard disks to record to. And given the relatively cheap price per gigabyte of hard disks these days, most people should be able to afford to buy enough hard disk space for their needs – without having to resort to destructive recording.

Non-destructive editing simply rearranges the order in which parts of the original audio are played back, skipping some parts completely and repeating others as required. These parts or bits of the original audio are referred to as 'regions' within the audio file. The default situation is for a region to encompass the whole of the audio file. You then define smaller regions within that file so you can repeat these, use them elsewhere in your session, or whatever. These regions are always shown in the Regions List that can be revealed or hidden at the right-hand side of the Pro Tools Edit window and can be used as many times as you like in your session without them using up any more hard disk space.

Pro Tools tracks can be regarded as 'playlists' or 'edit decision lists', because they typically contain one or more regions placed along the timeline. In other words, the track lists the regions in the order that you wish them to play back. Or you could say that the track lists the edits that you have made to your recordings so that they play back the way you want them to. When the audio is presented as a waveform that scrolls from left to right along a 'timeline' as it plays back, it is often referred to as a playlist. In Pro Tools, MIDI data can also be presented alphanumerically in a vertically scrolling list called the MIDI Event List. Audio is presented this way in many other digital editors, in which case the list is typically referred to as the Edit Decision List or EDL. Pro Tools does not have an alphanumeric audio event list or EDL – an unfortunate omission in my opinion as it can be far easier and quicker to make certain types of edits by typing alphanumerically into an EDL.

Audio Regions in Pro Tools can be displayed in several ways in the tracks, depending on what suits your purpose. The default is to show a representation of the analogue waveform ready for you to edit this, but you can simply display regions as blocks, for example, which can sometimes be less confusing to move around onscreen. The default display for MIDI Regions is to show the notes graphically, ready for you to edit these.

Non-real-time Digital Signal Processing can be applied to audio regions in Pro Tools by selecting a region and choosing a processor from the AudioSuite menu. Here you can reverse the audio region (try this for a Jimi Hendrix effect on your next guitar solo), or shift the pitch (try this on voices for 'Chipmunk' or 'Darth Vader' effects). Or you can change the level, apply EQ, compression, reverb, delay, modulation, or other effects.

Basic Editing

There are several basic concepts that you need to be familiar with as a Pro Tools editor. Let's start with the most basic editing commands.

Cut, Copy, Paste, and Clear Commands

Pro Tools has some similarities with word-processing and graphics software. For example, the way that the computer expects you to work is to select something first and then to issue a command that says what you want to do with whatever you have selected – either using keyboard commands or using the mouse, pointer, and menus. Once you have selected a region, for example, you can use the Cut, Copy, Paste, and Clear commands to rearrange and edit the material in your tracks.

You can select a region or regions using the Time Grabber tool or you can select a range along a track using the Selector tool. You can also work across multiple tracks. You can then use the Cut command to remove whatever you have selected from the Edit window and put this selection into the Clipboard ready to paste it elsewhere. The Clipboard is a temporary storage area in the computer's RAM (random access memory). You can use the Copy command to copy your selection into the Clipboard ready to paste elsewhere, without removing the original selection from the Edit window. You can use the Paste command to put the contents of the Clipboard into the Edit window at the Edit insertion point, overwriting any material that is already there. If you simply want to remove your selection without putting it into the Clipboard, use the Clear command instead.

Special Cut, Copy, Paste, and Clear Commands

Pro Tools also has four 'special' Edit menu commands that you can use for editing automation playlists (e.g. volume, pan, mute, or plug-in automation) on audio, Auxiliary Input, Master Fader, and Instrument tracks.

Cut Special, Copy Special, and Clear Special each have three sub-menu selections. These let you edit all the Automation or just the Pan Automation (whether the data is showing in the track or not), or just the Plug-in Automation on its own (when this is showing in the track).

The Paste Special command also has three sub-menu selections, the most useful of which is the 'Repeat to Fill Selection' command. This allows you to automatically fill a selection with audio regions or data much more quickly than by manually duplicating the regions.

tip ▷

To use the 'Repeat to Fill Selection', simply cut or copy an audio region so that it is in the Pro Tools software's Clipboard (i.e. temporarily stored into RAM), then make a selection in the Edit window and use the command to fill this selection. The Batch Fades window automatically opens to let you apply crossfades between the pasted regions.

Making an Edit Selection using the Edit Selection Indicators

You can use the Edit Selection indicators at the top of the Edit window to specify Edit selections numerically. Let's see how to do this:

Step 1. The Edit Selection indicators use whichever time format you have selected for the Main Time Scale, so if you want to work with Bars and Beats, you should make sure that you have selected this first.

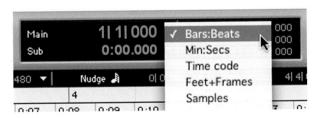

Fig. 9.1 – Setting the time format for the Main Time Scale Counter.

Step 2. Click with the Selector tool in the track you want to select.

Step 3. Click in the Start field at the top of the Edit window.

Step 4. Type in the start point for the selection and press the Forward Slash key (/) to enter the value and automatically move to the end field.

Step 5. Type in the end point for the selection.

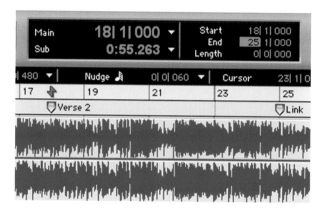

Fig. 9.2 – Edit window showing the End point (Bar 25) typed into the Edit Selection Indicators' End field.

Step 6. Press Enter to accept the value and you will see the Edit Selection made.

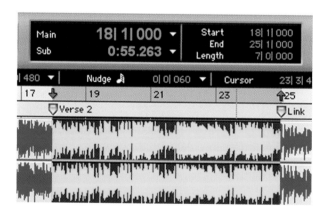

Fig. 9.3 – Edit window showing the Edit Selection made using the Edit Selection Indicators.

tip ▷ The above method can also be used to enter start and end values in the Transport window.

Working with Multiple Tracks

To make edits across multiple tracks or all tracks, you must select the tracks first. The easiest way to do this is to drag the insertion cursor across the tracks you want to select, dragging upwards or downwards to include adjacent tracks and dragging horizontally to define the time range.

Step 1. Choose the Selector tool.

Step 2. Point your mouse at any track of interest in the Edit window.

Step 3. Click and hold the mouse button, then drag horizontally to select a region.

Step 4. Without letting go of the mouse button, drag vertically upwards or downwards to select additional tracks of whatever type.

Step 5. Once you have made your selection, you can let go of the mouse button.

After you have made a selection and let go of the mouse button, if you change your mind and want to extend the selection to include more tracks or to shorten or lengthen the selection, press and hold the Shift key while you drag using the Selector tool. If you click in the Edit window without holding the Shift key, you will lose your original selection.

To select and edit all the tracks simultaneously, you can either drag with the Selector tool in any of the Timebase rulers with the Timeline and Edit Selections linked, or you can enable the 'All' edit group.

If you expect that you will be making the same edits to several tracks at once, you can group these tracks together so that edits applied to one will apply to all. Typically you might do this with your drum tracks, or with a brass section or a set of guitar tracks.

note ▷ If a group of tracks in the Edit window contains some tracks that are hidden (i.e. de-selected in the Tracks show/hide list at the left of the Edit window), these will not be affected by any edits made to the members of the group that are visible in the Edit window.

To paste to multiple tracks, engage the Selector tool then place the insertion point in each of the destination tracks by Shift-clicking in them. If you want to place the insertion point into all tracks, click in any of the Timebase rulers.

note ▷ Whenever you paste multiple types of data, whatever data has been copied is automatically pasted into the correct type of playlist.

Master Views

When you are working with data from multiple tracks, there are several things to keep in mind. For instance, if any of the selected tracks is set to a Master View, any edits you make will not only affect the audio or MIDI data on the track, they will also apply to any automation or controller data on the track.

So, when you are in a Master View, cutting an audio region also cuts any volume, pan, mute, send, or plug-in automation that is also on the track – saving you from having to individually cut data from each automation playlist on the track. Similarly, when an Auxiliary Input or Master Fader track is displayed in its Master View, any edits performed apply to all automation data in the track.

note ▷ When Audio tracks are set to display either Waveform or Blocks, they are said to be in a Master View. When MIDI or Instrument tracks are set to display Regions, Blocks, or Notes (when using the Selector tool), these tracks are said to be in a Master View. When Auxiliary inputs are set to display Volume, this is regarded as the Master View, and Master Fader tracks (which can only display Volume) are always considered to be in Master View.

Other Track Views

In any other track view with automation data displayed on the selected tracks, edits only affect the type of automation data displayed in each track.

For example, if track 1 displays Pan automation, track 2 displays Volume automation, and track 3 displays Mute automation, the Cut command cuts only pan data from track 1, volume data from track 2, and mute data from track 3.

tip ▷ Pro Tools lets you override this behaviour temporarily by pressing and holding the Control (Macintosh) or the Start key (Windows) while you choose the Cut, Copy, or Paste commands, enabling you to copy all types of automation on all selected tracks.

The Edit modes

Pro Tools has four Edit modes – Slip, Shuffle, Spot, and Grid – that can be selected by clicking the corresponding button in the upper left of the Edit window. You can also use the function keys on your computer keyboard, F1 (Shuffle), F2 (Slip), F3 (Spot), and F4 (Grid), to select the mode. The Edit mode that you have chosen affects the ways that regions may be moved or placed, how commands like Copy and Paste work, and how the various Edit tools (Trim, Selector, Grabber, and Pencil) work.

Slip Mode

The default mode that you should normally work in is the Slip mode. It is called Slip mode because it lets you move regions freely within a track or to other tracks and allows you to place a region so that there is space between it and other regions in a track.

In Slip mode, if you use the Cut command to remove a selection before the end of the last region on a track, it leaves an empty space where the data was removed from the track. You should also be aware that regions are allowed to overlap or completely cover other regions in Slip mode.

Shuffle Mode

In Shuffle mode, if you place two or more regions into a track, they will automatically snap together with no gap in between.

If you have an existing track that contains two or more regions with gaps between them, you can close the gaps by selecting Shuffle mode and using the Time Grabber tool to push a region in the direction of the previous region.

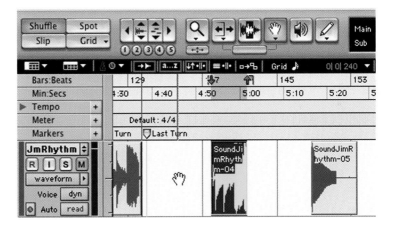

Fig. 9.4 – Selecting and moving a region in Shuffle mode.

As soon as you let go of the mouse, the gap between the two regions is closed, leaving them 'stuck together'.

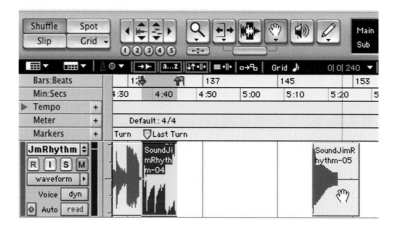

Fig. 9.5 – The region that you moved automatically attaches itself to the previous region in Shuffle mode.

So, if you want a region that you are moving around in the Edit window to automatically butt up against the previous region, with no overlap and with not even the smallest gap between them, you can use Shuffle mode. If you are in Slip Mode or even in Grid Mode you have to be careful to avoid overlapping the regions, and you will have to zoom in to see what you are doing. Consequently, placing regions accurately in these modes takes more time.

By the way, it is called 'Shuffle' mode because if you use the grabber to drag a region placed earlier in a playlist to a later position (or vice versa), the other regions will shuffle (i.e. move) their positions around to accommodate this re-positioning of the region. Similarly, if you use the Cut command to remove a selection before the end of the last region on a track, the regions to the right of the cut move to the left, closing the gap. Also, if you paste data anywhere before the end of the last region on a track, all regions beyond the insertion point move to the right to make room for the pasted material.

tip ▷ Be careful to return to Slip or Grid mode as soon as you have made your moves in Shuffle mode – it is all too easy to accidentally move a region and have Shuffle mode shuffle your regions to somewhere they shouldn't be. And if you don't notice this at the time it happens you may not be able to use even the multiple Undo feature to get back to where you were.

Spot Mode

Spot mode was originally designed for working to picture, where you often need to 'spot' a sound effect or a music cue to a particular Society of Motion Picture and Television Engineers (SMPTE) time code location.

The way this works is that you select Spot mode and then click on any region in the Edit window, or drag a region from the Region List or from a DigiBase browser into the Edit Window. The Spot dialog comes up and you can either type in the location you want or use the region's time stamp locations for spotting.

There are actually two time stamps that are saved with every region. When you originally record a region it is permanently time-stamped relative to the SMPTE start time specified for the session. Each region can also have a User Time Stamp that can be altered whenever you like using the Time Stamp Selected command in the Regions List popup menu. If you have not specifically set a User Time Stamp, the Original Time Stamp location will be set here as the default.

note ▷ If you have the optional DV Toolkit for Pro Tools LE installed, you can capture an incoming Time Code address and spot the region to this.

tip ▷ Spot mode can also be very useful when editing music projects in Pro Tools – particularly if you move a region out of place accidentally. Using Spot Mode you can always return a region to where it was originally recorded. Also, as long as you remember to set a User Time Stamp if you rearrange regions to locations other than where they were first recorded, you can always return regions to these locations.

For example, if you move a region by accident, using Spot Mode would be an ideal way to put it back to exactly where it came from.

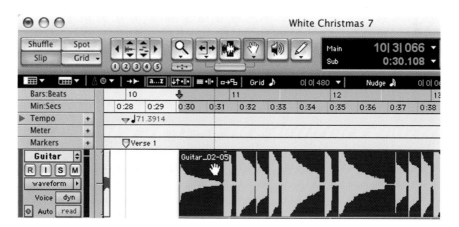

Fig. 9.6 – Edit window showing a Region accidentally moved from its original position.

Step 1. Put the software into Spot mode by clicking on the Spot mode icon at the top left of the Edit window.

Step 2. Select the Grabber tool.

Step 3. Click on the region to bring up the Spot dialog.

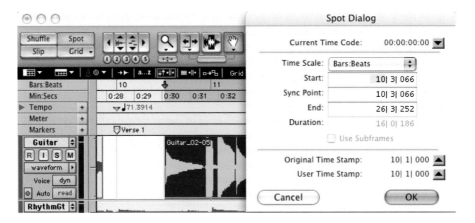

Fig. 9.7 – Using the Spot Dialog to return a Region to its original position.

Step 4. If the Region has been moved from the position at which it was originally recorded, the current Start position displayed in the Spot Dialog will be different from the Original Time Stamp position shown in the dialog. You can either type the correct Start position or (even quicker) just click the upwards-pointing arrow next to the Original Time Stamp to put this value into the Start position field.

Fig. 9.8 – Click the Original Time Stamp arrow to enter this location into the Start field in the Spot Dialog.

note ▶ If you have deliberately moved the Region since recording it and have taken the trouble to enter a User Time Stamp in the Regions List for the moved region, you should enter the User Time Stamp into the Start field instead.

Step 5. When you click 'OK' in the Spot Dialog, the region will be moved back to the location where it was originally recorded.

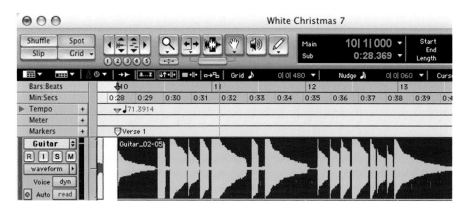

Fig. 9.9 – Edit window showing Guitar Region back in its original position.

Grid Mode

Grid mode lets you constrain your edit selection to gridlines that correspond to a grid value that you can choose to suit your purpose. Grid mode is particularly useful if you are editing pattern-based music that starts and ends cleanly at regular boundaries, such as bars or beats. You can choose the Grid size using the Grid value popup menu located above the Timebase Rulers and the tracks in the Edit window.

tip ▷ If you press and hold the Control and Option keys (Mac) or the Start and Alt keys (in Windows) you can use the plus (+) and minus (−) keys on the numeric keypad to increment or decrement the Grid size.

The Grid size can be based on a time value using the Main Time Scale; or, if Follow Main Time Scale is deselected, another time format can be used for the Grid size. To make the Grid lines visible in the Edit window, just click on the currently selected Timebase ruler name (the one highlighted in blue) to toggle these on and off.

tip ▷ You can temporarily suspend Grid mode and switch to Slip mode by holding down the Command key (the Control key in Windows), which is very useful while you are trimming regions, for example.

Grid mode can be applied either in an Absolute or Relative way. In Absolute Grid mode, regions are 'snapped' onto Grid boundaries when you move

239

them – so regions can never be placed in between the currently applicable Grid boundaries. This is what you normally expect and want Grid mode to do and is the default behaviour.

Relative Grid Mode lets you edit regions that are not aligned with Grid boundaries as though they were. For example, in 4/4 time signature, if a region's start point falls between beats and the Grid is set to 1/4 notes, dragging the region in Relative Grid mode will preserve the region's position relative to the nearest beat.

In a music recording, for example, a musician may have deliberately played a note just before the beginning of a new bar. If you separate this note into its own region, then use the Grabber tool to move it earlier or later by an exact number of beats using the Absolute Grid mode, the region's start position will be 'snapped' exactly onto the new beat position, and will not sound as the musician intended it to.

If you choose Relative Grid mode and set a suitable Grid value that allows sufficient time before the bar, then when you move the region containing the note along the Grid, it 'snaps' to positions that preserve the note's original positioning just a little before the beat.

To select Relative Grid mode, click and hold the Grid button, move the mouse to where it says 'Relative Grid', then let go.

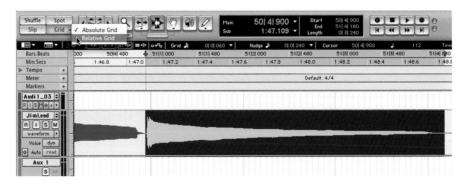

Fig. 9.10 – Selecting Relative Grid Mode.

As you can see from the screenshot below, the guitar note is played just a little before the beginning of Bar 51 in this example – 42 ticks earlier, to be precise.

The Grid is set, in this example, to 60 ticks, representing a 64th note. Notice that this is just a little more than 42 ticks before Bar 51, so you can see a blue grid line a little to the left of the region.

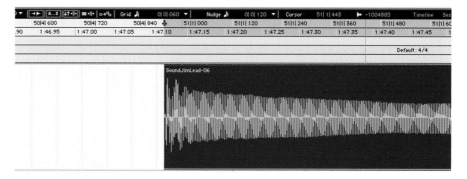

Fig. 9.11 – Note positioned 42 ticks before Bar 51.

With Relative Grid mode selected, when you use the Grabber tool to slide the region to right or left along the Timeline, the region snaps to start positions that are exactly 42 ticks (in this example) before the new bar position, thus preserving the note's original positioning relative to the bar positions.

In the example below, I dragged the region earlier by one 64th note grid position in Relative Grid Mode and as you can see from the screenshot, it is now positioned a little (exactly 42 ticks in this example) before the last 64th note of Bar 50.

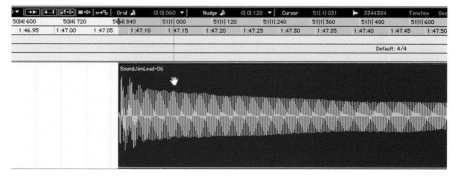

Fig. 9.12 – Note dragged back along the Timeline in Relative Grid mode to a position exactly one 64th note earlier.

Dragging the region in Absolute Grid mode would have only allowed me to position the start of the region containing the note exactly onto a 64th note. When I do this, as you will see from the accompanying screenshot, the note starts exactly on a 64th note grid position – which in this instance would sound wrong.

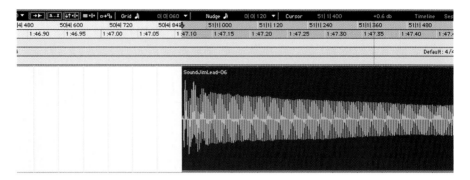

Fig. 9.13 – Note dragged back along the Timeline in Absolute Grid mode to a 64th note grid position.

Making Timeline Selections

To make a Timeline selection (as opposed to an Edit selection among the tracks), you can either type your selection into the Transport window or drag with the Selector or Time Grabber tool in any Timebase ruler.

The Timeline selection is indicated in the Main Timebase ruler by blue Playback Markers, which turn red if a track is record-enabled. The start, end, and length for the Timeline selection are displayed in the corresponding fields in the Transport window.

You can also set the Timeline selection by dragging the Playback Markers. Using either the Selector or Time Grabber tool, drag the Start Playback Marker (the down arrow) to set the start point and drag the End Playback Marker (the up arrow) to set the end point.

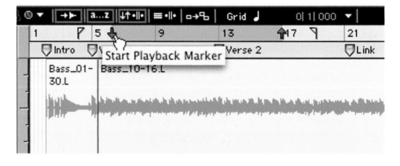

Fig. 9.14 – Setting the Timeline Selection by dragging the Playback Markers.

tip ▶ If you want to constrain the selection to the current Grid value, set the Edit mode to Grid.

If you press the Option key (Mac) or the Alt key (Windows), you can drag either the Start or End Playback Marker and the whole Timeline selection will move forwards or backwards along the Main Timebase ruler as you drag.

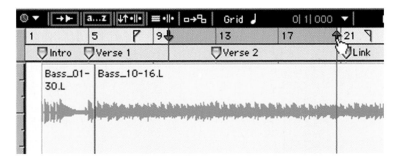

Fig. 9.15 – Dragging the Timeline Selection while holding the Option (Alt) key to move the whole Timeline selection.

If the Timeline and Edit selection are unlinked, you can drag either the Start or End Edit Marker (not the Playback markers) while holding the Option (Alt) key and the whole Edit selection will move backwards or forwards along the Timeline as you drag.

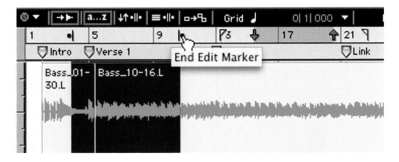

Fig. 9.16 – Dragging the Edit Selection while holding the Option (Alt) key to move the whole Edit selection.

Linked Timeline and Edit Selections

With the Timeline and Edit selections linked, any time you make an Edit selection in a track or tracks, a corresponding selection is automatically made in the Timebase Rulers. This is the default behaviour.

To link the Timeline and Edit selections, select Link Timeline and Edit Selection from the Options menu or click the Link Timeline and Edit Selection button in the Edit window so that it becomes selected (outlined in blue).

Fig. 9.17 – The Link Timeline and Edit Selection button.

With the Timeline and Edit selections linked, when you make an edit selection in a track the same time range automatically becomes selected in the Timebase Rulers and a small blue arrow is placed at each side of the selected range.

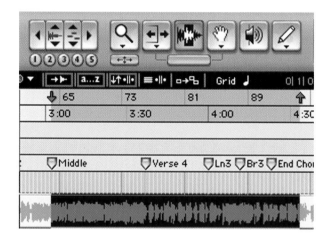

Fig. 9.18 – Edit window showing a highlighted Edit selection in a track with the corresponding Timeline selection made in the Timebase rulers.

tip ▷ Press Shift-Forward slash (/) to toggle 'Link Timeline and Edit Selection' on and off.

Unlinked Timeline and Edit Selections

When you unlink the Timeline and Edit selections, you can make different selections in the Edit display and in the Timeline above the Edit display.

To make a Timeline Selection, choose the Selector tool and drag the cursor along the Main Timebase Ruler to set the playback and recording range.

tip ▷ You can edit this selection by dragging the blue arrows that appear in the Timebase Ruler to mark the beginning and end of your Timeline selection.

You can enter the Start and End locations into the Edit Selection indicators at the top of the Edit window, or you can choose the Selector tool and drag the cursor across the Edit window to make an Edit selection.

tip ▷ You can edit this selection by dragging the black brackets that appear in the Timebase Ruler to mark the beginning and end of your Edit selection.

So why would you want to unlink the Timeline and Edit selections? Take a look at the screenshot below in which the Timeline selection is defining the range to be looped during playback while a MIDI region within the loop is selected for editing. During playback, the Edit selection can be nudged, quantized, or transposed while the loop plays back without interruption – which is just fine and dandy!

Fig. 9.19 – Timeline selection defines the playback loop while the Edit selection in the MIDI track can be moved or otherwise edited during playback.

Here's another example. Say you are working to picture and you want to find or audition some audio material that is at a completely different location than the current Timeline selection. Again, you will find it extremely useful to be able to unlink the Timeline and Edit selections. Take a look at the screenshot below to see the Edit window with completely different Timeline and Edit selections.

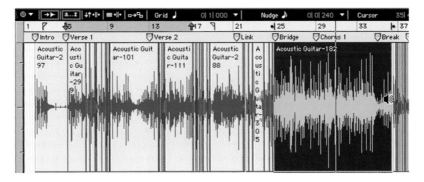

Fig. 9.20 – Different Timeline and Edit selections in the Edit window.

In the Edit menu there is a command to Play Edit Selection instead of playing the Timeline Selection (which is what happens when you use the normal Play command or hit the Spacebar). The current Timeline Selection stays in place when you play the Edit selection using this command, so you can search through the Edit window, selecting and auditioning regions of potential interest until you find material that you like. Then you can then go back to the Timeline selection and place the material that you have found within the context of the scene.

The Playback and Edit Markers

Timeline selections are indicated in the Main Timebase ruler by Playback Markers. These appear as blue arrows normally and as red arrows when any track is record enabled.

If you have entered and enabled Pre- or Post-Roll amounts in the Transport window, these will be indicated in the Main Timebase Ruler by a pair of green flags. If you disable the Pre- and Post-Roll amounts, these will still be indicated, but the flags will be coloured white.

Take a look at the screenshot below to see how the Edit window might look with the Timeline and Edit selections linked and with Pre- and Post-Roll amounts set and enabled. Notice that the Edit selection is represented in the Main Timebase Ruler by the blue Playback Markers that also indicate the Timeline Selection. In this example the Main Timebase Ruler is set to Bars:Beats.

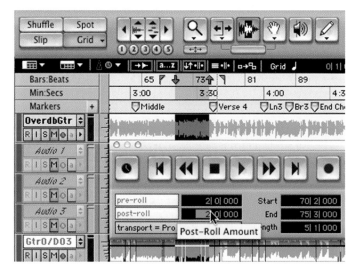

Fig. 9.21 – Blue Playback Markers and Green Pre- and Post-Roll flags indicate the Timeline Selection and the Pre- and Post-Roll Selection, respectively.

Take a look at the screenshot below to see how the Edit window might look with the Timeline and Edit selections unlinked and different selections made in the Timebase Ruler and waveform display. When the Timeline and Edit selections are unlinked, Edit selections are displayed in the ruler with Edit Markers, which appear as black brackets.

In this example, a track's Record Enable button has been engaged so that we can see how the Playback Markers look in red. At the bottom right, you can see the Edit selection in the waveform display and the two black brackets in the Main Timebase ruler above this. In the middle of the screenshot you can see the Timeline selection indicated by the red arrows in the Main Timebase ruler.

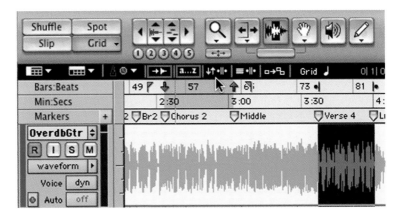

Fig. 9.22 – Red Playback Markers indicate the Timeline Selection while black brackets indicate where the Edit Selection is.

Linked Track and Edit Selections

With the Track and Edit selections linked, if you make an edit selection within a track or across multiple tracks, each associated track will automatically become selected, with the track names highlighted.

To link Track and Timeline selections, select Link Track and Edit Selection from the Options menu or click the Link Track and Edit Selection button near the upper left of the Edit window so that it becomes outlined in blue.

In the screenshot below, you can see that the 'Link Track and Edit Selection' button is active, and an Edit Selection has been made in the Edit window's waveform display. This has caused the track names at the left of the Edit window to become highlighted, indicating that these tracks have become selected.

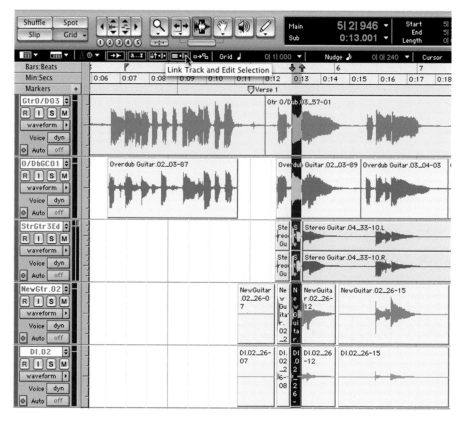

Fig. 9.23 – Edit window with Link Track and Edit Selection active, with an Edit selection just made in the waveform display, and with the track names at the left highlighted to indicate that these have just been automatically selected.

So why might you want to do this? Well, it can be very useful to have the tracks selected when you are editing waveforms and automation data so that you can quickly apply track-level commands such as Track View toggle, or change track heights, and have the command apply to all tracks you are working on.

Finding your way around the Edit window

If you are new to Pro Tools you will have to spend some time getting used to working in the Edit window. You will need to be comfortable and familiar with several keyboard commands and menu items and with the buttons at the top of the Edit window. You will need to know how to loop playback, how to control screen scrolling, and lots more stuff like this. So let's get started with some of this stuff right now!

Track View Toggle

If you press Control-minus on the Mac (Start-minus in Windows) on your computer keyboard (using the minus (−) symbol on the QWERTY keyboard, not the minus symbol on the numeric keyboard found at the right-hand side of some keyboards), this toggles the Track View between Waveform and Volume.

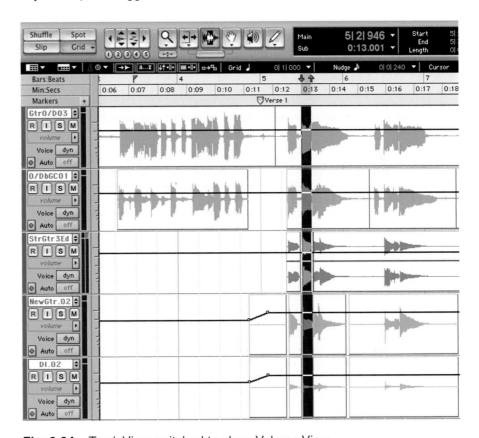

Fig. 9.24 – Track View switched to show Volume View.

Zoom Toggle

The Zoom Toggle button in the Edit window lets you define a zoom state (with separate track heights for audio and MIDI track views) and toggle between it and the current zoom state.

When the Zoom Toggle button is enabled, the Edit window switches to display the stored zoom state. If you then make any changes to the Zoom Toggle parameters, these will automatically be stored in the zoom state. When Zoom Toggle is disabled, the Edit window reverts to the last zoom state.

To set up the Zoom Toggle state:

Step 1. Click on the Zoom Toggle button to select and highlight it.

Step 2. Adjust the Zoom Toggle parameters, that is, the Track Heights, the Edit window view (e.g. Waveform, Volume, or other automation view), the Grid settings, and the Audio and MIDI zoom levels.

Step 3. When you have chosen settings for the Zoom Toggle state that suit the type of edits you want to make, click the Zoom Toggle button once more to de-select it and return to your previous zoom settings.

Step 4. Now, when you make another edit selection, you can simply click the Zoom Toggle button to recall the stored zoom state.

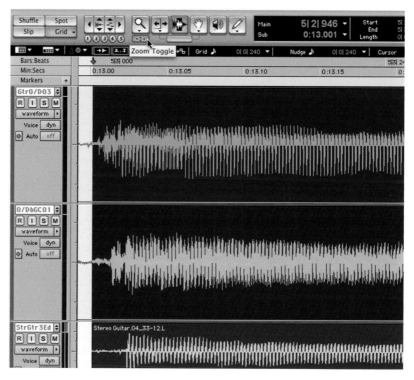

Fig. 9.25 – Zoom Toggle button enabled to switch the display to the stored Zoom Toggle state.

Getting used to Zooming around

Pro Tools lets you zoom around in the display in various ways, which you can choose according to the style you prefer. You can click on the Zoom buttons to zoom vertically or horizontally, or you can use the Zoom tool.

The Zoom Tool lets you zoom in and out around a particular area within a track using either its Normal Zoom mode or the Single Zoom mode.

To zoom horizontally in Normal Zoom mode, just drag along the track with the Zoom tool selected. To zoom both horizontally and vertically at the same time, press the Command key (Mac) or Control key (Windows) while you drag.

The Single Zoom mode automatically re-selects the tool that you were using previously after it zooms the display.

If you press and hold the Command and Control keys (Mac) or Control and Alt keys (Windows) then move your mouse cursor into the Ruler area in the Edit window, you can zoom in without having to select the Zoom tool first. If you just click once at any position in the ruler, the display zooms in one level, and if you drag across a range in the Ruler, the display will zoom in around this range.

Continuous Zooming

With the Zoom tool selected, you can zoom in or out continuously by holding the Control (Mac) or Start (Windows) modifier key on the computer keyboard while you drag vertically or horizontally in the Edit window. Drag up to zoom in vertically, drag down to zoom out vertically, drag to the right to zoom in horizontally, or drag to the left to zoom out horizontally.

For example, look at how the display changes between the two screenshots below as the Zoom tool is dragged horizontally to the right while holding the modifier key.

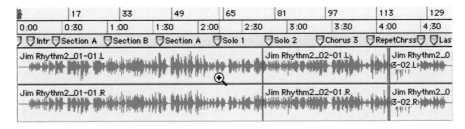

Fig. 9.26 – Before dragging the Zoom tool horizontally to the right while holding the modifier key.

Notice that the display zooms in smoothly and continuously as you drag to the right horizontally – so you see a smaller section of the waveform with a higher resolution of detail.

Notice also that all the tracks in your session are zoomed horizontally at the same time.

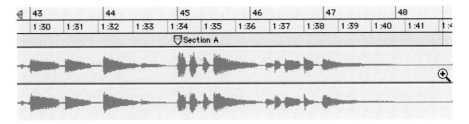

Fig. 9.27 – After dragging the Zoom tool horizontally to the right while holding the modifier key.

Now let's look at how the display changes as the Zoom tool is dragged vertically upwards while holding the modifier key.

tip ▷ If you want all the audio tracks to zoom vertically, press the Shift key as well as the Control (Mac) or Start (Windows) modifier key.

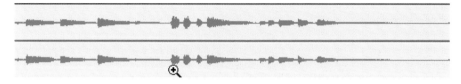

Fig. 9.28 – Before dragging the Zoom tool vertically upwards while holding the modifier key.

Notice that the display zooms in smoothly and continuously as you drag vertically upwards – making the waveform larger and easier to see.

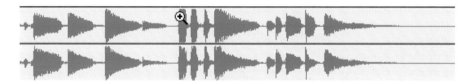

Fig. 9.29 – After dragging the Zoom tool vertically upwards while holding the modifier key.

tip ▷ When you have settled on a zoom level that you like to work with, you can save this into any of the five Zoom Preset buttons below the Zoom Tools. Set the zoom level first, then just click and hold any of the Zoom Preset buttons and choose Save Preset from the popup selector that appears.

Looped Playback

There are lots of times when you will want to loop playback during your editing sessions. For example, if you are editing a selection that you intend to repeat to form a larger section, you will find it very useful to loop this selection before you repeat it to make sure that it loops without any glitches. For instance, you might take a 4-bar 'groove' section and repeat this four times to form a 16-bar chorus section.

If you do hear a glitch at the loop point, this probably means that the length of the selection is not exactly right, so you should adjust the start and/or end of the selection until it does loop without glitching.

When Loop Playback is enabled in the Options menu, a loop symbol appears in the Play button in the Transport window and the selected range in the Edit window repeats when you playback. You can also enable Loop Playback by Control-clicking (Mac) or right-clicking (Windows) on the Play button in the Transport window.

tip ▷

If you select the range to loop using the Selector tool in the Edit window, then you should enable 'Link Timeline and Edit Selection' so that the same time range is automatically selected in the Edit window and in the Timebase Ruler.

Scrolling Options

There are just two scrolling options for the Edit window in Pro Tools LE and M-Powered: Scroll After Playback and Page Scroll. Scroll After Playback does what it says – not moving the display until you hit Stop. The Page Scroll option keeps the display stationary until playback reaches the right-most side of the page, then it quickly changes the display to show the next section (i.e. page) of the waveform that will fit within the Edit window. It keeps on doing this until you stop playback.

note ▷

The playback position jumps back to the position you started playing from when you stop or pause playback, unless you enable the 'Timeline Insertion Follows Playback' in the Operation Preferences window.

Timeline Insertion Follows Playback

With 'Timeline Insertion Follows Playback' enabled in the Operation Preferences dialog, when you press play, then stop or pause, then press play again, Pro Tools 'picks up where it left off.' In other words, when you stop, the insertion point 'parks itself' at that point – instead of returning to the position it was at when you first pressed play.

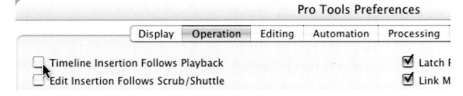

Fig. 9.30 – Setting the 'Timeline Insertion Follows Playback' Preference.

note ▷ There is a keyboard command to enable or disable this preference. Just press the 'n' key to toggle the Timeline Insertion Follows Playback preference on and off.

tip ▷ It is all too easy to press the 'n' key on your computer keyboard by accident. So if Pro Tools is not operating the way you want it to, just hit the 'n' key again to change this preference back to the way you want it.

Useful Shortcuts

You will probably want to adjust the track heights during your editing session and there are a couple of useful shortcuts that you should learn. To change all the track heights at the same time, simply hold the Option (Alt) key while you change any one of the track heights. A shortcut when you want to change one track or group of tracks is to put the insertion cursor into the track, hold the Control (Start) key, then press the Up or Down arrow keys to choose the track height.

To quickly switch audio tracks between Waveform and Volume view, hold down the Control key and press the minus ('−') key on the computer keyboard.

To scroll the Edit (or Mix) window one 'page' or 'screen' to the right or the left, hold the Option (Alt) key and press the Page Up or Page Down buttons on your computer keyboard. On the Mac, these keys have an upwards or downwards pointing arrow with two short horizontal lines across the stem of the arrow.

You can also toggle the display to simultaneously increase the height of the track to Large and to zoom the display to encompass the selection by clicking Control-E (Mac) or Start-E (Windows). Take a look at the screenshots below to see how this works.

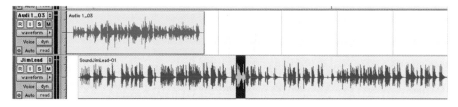

Fig. 9.31 – Edit window showing two tracks set to Medium height and with an Edit selection.

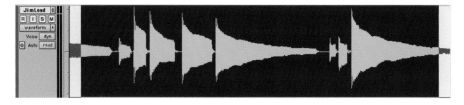

Fig. 9.32 – Edit window after switching the display to make the track height Large and to encompass the Edit selection.

Finding the ends of the selection

If you have made a selection in the Edit window and then you zoom in to fine-tune your edit points, the display will zoom in around the start point of the selection. If you want to jump to the end point, you can use the right arrow key on the computer keyboard to move the display so that the right-hand edge of the selection, that is, the end point, is in the centre of the screen. To get back to the start point of the selection, just hit the left arrow.

To see how this works:

Step 1. Make a selection in the Edit window then zoom the display in horizontally so that the end of the selection is no longer visible.

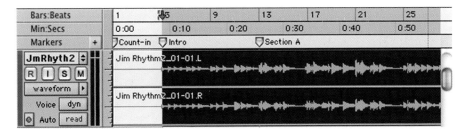

Fig. 9.33 – Edit window showing the start point of the Edit selection.

Step 2. Press the right arrow key on the computer keyboard. The end point of the selection is moved to the centre of the Edit window.

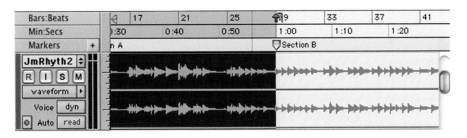

Fig. 9.34 – Edit window showing the end point of the Edit selection centred in the Edit window.

Step 3. Press the left arrow key to get back to the start point of the selection again.

Working with Regions

Regions are the basic building blocks for arranging audio and MIDI in Pro Tools so you need to make sure that you know how these are created, edited, and arranged. The Pro Tools Reference Guide covers this thoroughly, so I recommend that you keep this to hand during your editing sessions until you become completely familiar with the software.

When you record new audio or import existing audio files, Pro Tools creates a Region that plays back the entire file when placed into a playlist on a track. Very often you will use the Trimmer tool to remove some audio from the start or end of the recording. This creates a new Region in the playlist that is shorter than the original region, and this new region also appears in the Region List.

In fact, new regions are often created automatically when you make edits to existing regions. For example, if you use the Clear command to remove a section from within a region, the sections on either side of this form new regions. If you use the Cut command instead, the section you are cutting is placed onto the Clipboard and a new region representing this section is also added to the Region List.

The Region menu contains commands that you will frequently apply to selected regions, such as Mute/Unmute, Lock/Unlock, Group and Ungroup, Capture. . . and Quantize to Grid, most of which are fairly self-explanatory.

The Capture. . . region command defines a selection as a new region and adds it to the Region List. From there, the new region can be dragged to any existing tracks.

You can also create new regions using the Separation Grabber tool. Using the Separation Grabber saves you the trouble of separating the region first.

Make a selection using the Selector tool. Then select the Separation Grabber tool from the popup selector that appears when you press and hold the Grabber tool's button. Take a look at the screenshot below to see how this might look.

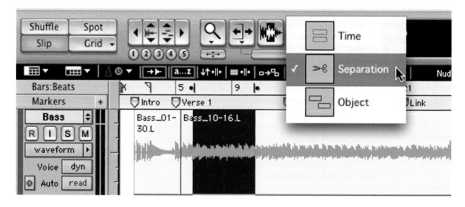

Fig. 9.35 – Choosing the Separation Grabber tool.

When you have selected the Separation Grabber tool, you will see a small pair of scissors appear on the 'hand' to remind you that the Separation Grabber is currently selected. Now you can drag your selection to a new location within the same track or on another track. The selection is automatically separated from its 'parent' region and a new region containing this selection is created. As usual, new regions are also created from the material outside the original selection. Take a look at the screenshot below to see how this might look.

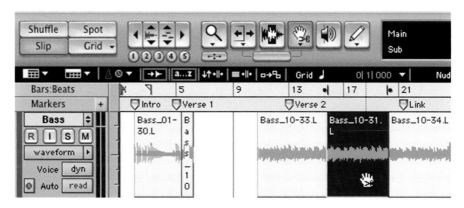

Fig. 9.36 – Using the Separation Grabber tool to separate a selection from a region and move this new region to a new location.

tip ▷ If you want to leave the original region intact, press the Option key (Mac) or the Alt key (Windows) while you drag the selection to the new location. A new region containing the previous selection is created and placed at the new location. The original region remains intact.

The Edit menu also has commands that affect regions, such as Separate Region, Heal Separation, and various useful Trim Region commands.

There are actually three Separate Region commands. If you have made an Edit selection using the Selector tool you can separate this to form a new region using the 'At Selection' command from the Separate Region sub-menu. Or you can type Command-E (Mac) or Control-E (Windows) on your computer keyboard to invoke the 'Separate Region At Selection' command if you prefer.

tip ▷ If you simply insert the Selection cursor into the waveform without making an Edit selection, the 'Separate Region At Selection' command will split the region at this insertion point instead.

257

If you need to separate a region into lots of smaller regions, you can use one or other of the two additional choices available in the Separate Region sub-menu: 'Separate Regions On Grid' separates regions based on the currently displayed Grid values and boundaries. 'Separate Regions At Transients' separates regions at each detected transient.

note ▷ With looped regions, the Separate Regions commands automatically unloop and flatten the looped regions before separating them.

The Heal Separation command returns separated regions to their original state – provided that the regions are still next to each other and that their relative start/end points haven't changed since they were separated.

Once you have created a region, it appears in the Region List. From the Region List you can drag it to a track to add to an existing arrangement of regions or to create a new arrangement 'from scratch'. You can slide regions or groups of regions around freely in the Edit window using the Time Grabber tool in Slip mode. You can also move regions around while constrained to the grid in Grid mode, or shuffle them around in Shuffle mode. You can also 'spot' regions to exact locations using Spot mode, which is particularly useful in post-production.

Sync Points

When you are placing regions in Spot mode or Grid mode, it is sometimes useful to align a particular point within a region with a specific Timeline location – instead of aligning the start point of the region, which is the default situation. To cater for such situations, Pro Tools lets you define a sync point for any region.

This situation often comes up when you are laying up music and sound effects to picture. The standard example quoted here is where you have a creaking door that eventually slams shut. The sound effect file includes the creak and the slam, and the obvious thing to do here is to line up the sound of the door slamming shut with the video frame at which the door actually slams shut – which is some way into the audio region representing this sound effect.

You can define a region's sync point in Slip mode by first placing the Selector tool's insertion cursor at exactly the point within the region where the sound that you want to synchronize to starts. Then choose the Identify Sync Point command from the Region menu to identify this as the sync point for the region.

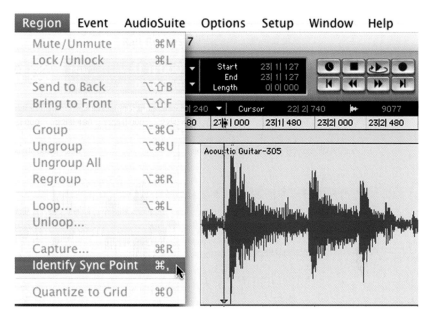

Fig. 9.37 – Setting a sync point within a region.

A small down arrow appears at the bottom of the region, with a vertical, light grey line above this indicating the location of the sync point. You can move this sync point to anywhere else inside the region using the Time Grabber tool to drag it earlier or later.

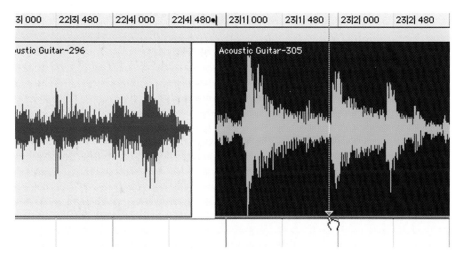

Fig. 9.38 – Dragging a sync point to a new position within the region using the Time Grabber.

Hearing the audio scrubbing back and forth can often be far more revealing than looking at the waveform when you are trying to identify what sounds best. So you may find it helpful to use the Scrubber tool to help you locate where the sync point should be.

To remove a sync point, just Option-click (Mac) or Alt-click (Windows) on the sync point using the Time Grabber or Scrubber tools.

Nudging Regions

Pro Tools lets you nudge regions (or MIDI notes) by small increments or decrements along the tracks in the Edit window. The way this works is that you set a Nudge value using the Nudge Value popup menu, then select a region or group of regions; then you move (i.e. nudge) these forwards or backwards along the Timeline by pressing the plus (+) or minus (−) keys on the numeric keypad.

You can nudge material while Pro Tools is playing back, which really helps when you are fine-tuning 'grooves'. You can nudge continuously in real time to adjust the timing relationship between tracks and you can even nudge the positions of automation breakpoints in the playlists.

The Nudge value not only determines how far regions and selections are moved when you press the nudge keys, it can also be used to move the Start and end points for selections by the Nudge value or to trim regions by the Nudge value.

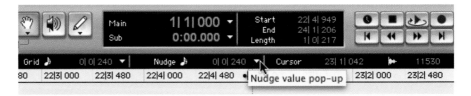

Fig. 9.39 – Setting the Nudge value.

A popup selector near the top of the Edit window lets you choose the Nudge value. With the main counter set to Bars:Beats, for example, the values offered are the common sub-divisions of a bar. You can also type the values you want directly into the Nudge Value display, which is useful if the values you want to use are not listed in the popup.

tip ▷ If you press and hold the Command and Option keys (Mac) or the Control and Alt keys (in Windows) you can use the plus (+) and minus (−) keys on the numeric keypad to increment or decrement the nudge value.

Fig. 9.40 – The Nudge value popup.

tip ▷

You can also nudge a region's contents (sliding audio or MIDI into and out of the current region boundaries) while keeping the region's start and end points exactly where they are – assuming that there is material outside the region's start and end points that can be slid into or out of the region. Using the Time Grabber tool, select a region whose contents you want to nudge. Press and hold the Control key (Mac) or the Start key (Windows) while you use the plus (+) and minus (−) keys to nudge the contents of the region without changing the region's start and end points.

Useful Region Editing Commands

The Edit menu and the Region menu both contain useful commands that you can apply to regions.

Shift Region

You can use the Edit menu's Shift command to move track material forward or back in time by a specified amount. The Shift command can operate on selections, regions, MIDI notes, MIDI controller data, and automation breakpoints. Note that if you have selected just part of a region, when you shift it, new regions are created from the selection and from any material outside of the selection.

Fig. 9.41 – The Shift dialog.

261

Duplicate Region

The Edit menu's Duplicate command copies a selection and places it immediately after the end of the selection. Though this is similar to using Copy and Paste, Duplicate is more convenient and faster, particularly when working with data on multiple tracks.

tip ▷ To duplicate a region and automatically place it immediately before the selected region choose the Time Grabber tool then click on the region while holding all three modifier keys: Command-Option-Control (on the Mac) or Start-Alt-Control (in Windows). And it gets better! If you click on any other region in the track first to select it, then click on the region you want to duplicate while holding the three modifier keys, a duplicate of this region will attach itself immediately before the region you first selected.

Repeat Region

The Edit menu's Repeat command is similar to Duplicate, but allows you to specify the number of times the selected material is duplicated. The material is placed immediately after the selection's end point, and duplicated by the number of times specified.

note ▷ When using Duplicate (or Repeat) for audio that must fall cleanly on the beat (such as rhythmic loops), it is important that you select the audio material with the Selector tool, or by typing in the start and end points in the Event Edit area. If you select an audio region with the Time Grabber tool (or by double-clicking it with the Selector tool), the material may drift by several ticks because of sample-rounding.

If, on the other hand, you want to Duplicate (or Repeat) audio that is not bar- and beat-based, set the Time Scale to any format except Bars:Beats. This ensures that the duplicated audio material will have the correct number of samples and will be placed accordingly.

Lock Region

If you want to prevent a region from being moved or deleted you can use the Region menu's Lock command. A small lock icon appears in the region to warn you that it has been locked. Locking regions can help to prevent you from accidentally moving these – especially in Shuffle mode. The neat thing here is that in Shuffle mode, locked regions, and all regions occurring after the locked region, will stay exactly where they are even though regions on neighbouring tracks are being shuffled around.

Mute Region

The Region menu's Mute/Unmute command mutes playback of any selected regions and dims these to let you know that they are muted. Choose the command a second time to unmute the selected regions.

Quantize to Grid

The Region menu's Quantize to Grid command adjusts the placement of selected audio and MIDI regions so that their start points (or their sync points, if they contain one) line up exactly with the nearest Grid line. Choose a Grid value then use the Selector or Grabber tool to select the regions you want to quantize, making sure that the whole of each region is selected. When you choose the Quantize To Grid command, selected regions (or their sync points) are moved to the nearest Grid line.

Region Groups

Many small regions, such as the individual hits of a drum pattern, can be created using Beat Detective or the 'Separate Region At Transients' command, or can be imported as REX or ACID files. These can be unwieldy if you need to move them around while you are arranging your music, so Pro Tools provides a set of region commands that let you manipulate regions in groups. Region groups are also useful for grouping parts and sections together so that they can be easily copied and moved. So, for example, you might group all the backing vocals together in the first chorus so that you can copy and paste these into subsequent choruses.

note ▷

A region group is a collection of any combination of audio and MIDI regions that looks and acts like a single region. Region groups can be created on single or multiple adjacent audio, MIDI, and Instrument tracks and can include both tick- and sample-based tracks.

To create a region group, select one or more regions on one or more tracks and choose the Group command from the Region menu.

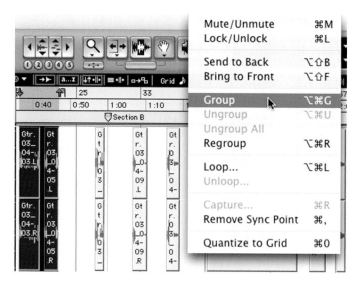

Fig. 9.42 – Using the Group command to group two regions.

The region group will appear as one region with the Region Group's icon in the lower left corner and will be added to the Region List.

You can 'nest' region groups together with other regions or region groups. Take a look at the screenshot below and observe the region group created in the example above about to be nested within a new region group comprising this region group and three more regions.

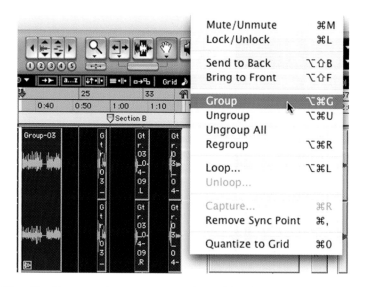

Fig. 9.43 – Nesting a region group together with other regions into a new region group.

The Region menu also has commands to let you ungroup and regroup regions. If you apply the 'Ungroup' command to nested region groups, it will only ungroup the top-layer region group, leaving any underlying region groups untouched. The 'Ungroup All' command, on the other hand, will ungroup a region group together with all of its nested region groups.

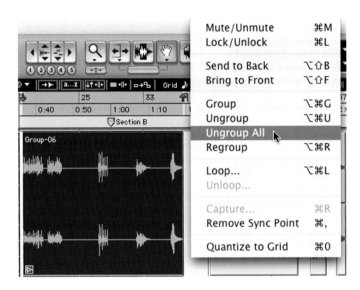

Fig. 9.44 – Using the Ungroup All Region command.

As you can see from the screenshot below, after using the 'Ungroup All' command, the grouped region is disassembled so that the original regions are accessible again.

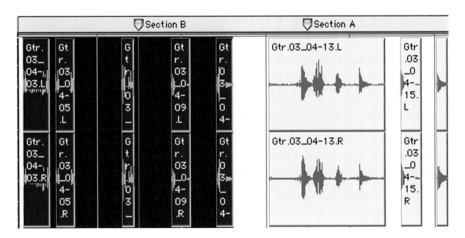

Fig. 9.45 – Regions all ungrouped.

With the regions ungrouped you can edit these individually, as necessary. As long as you don't group and ungroup any other regions beforehand, you can then use the Regroup command to regroup these as they were and continue working on your arrangement.

You can also create multitrack region groups by selecting regions on multiple adjacent tracks.

Step 1. Select regions across multiple adjacent tracks.

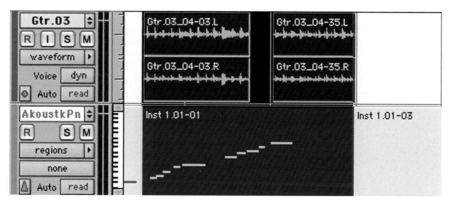

Fig. 9.46 – Preparing to create a mixed multitrack Region Group.

Step 2. Choose the Group command from the Region menu.

If you group a mix of sample- and tick-based audio or MIDI tracks together, a different region group icon in the bottom left corner of the region group indicates this. Take a look at the screenshot below to see how this looks.

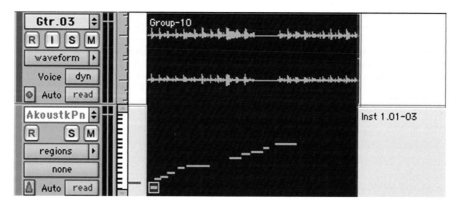

Fig. 9.47 – A mixed multitrack Region Group containing both sample-based and tick-based tracks with a mixed region group icon in the lower left corner.

note ▷ Region groups can become separated if you move a track from the group so that it is no longer adjacent, hide a track from the group, insert a track within the group, delete a track from the group, record into a region group, change the tempo of a mixed group, or change playlists on a track within the group. A broken icon appears in the lower left corner to warn you if this has happened.

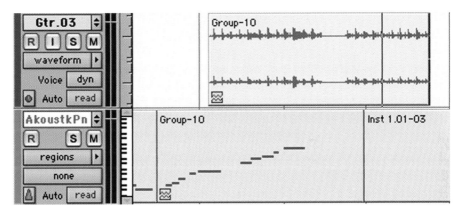

Fig. 9.48 – A mixed multitrack Region Group separated by changing tempo.

tip ▷ Pro Tools can export and import region groups using the new '.rgrp' region group file format. This can be very useful for creating multitrack loops containing references to all the audio files within the region group, region names and relative location in tracks, track names, fades and crossfades, and all the MIDI data from the region group. You can then import these into your current or future projects by dragging and dropping the region group file from a DigiBase browser or from Windows Explorer or Macintosh Finder to the Timeline, a track, the Track List, or the Region List.

note ▷ Region group files don't store automation, plug-ins, track routing, tempo or meter maps, or Region List information.

Region Looping

In common with most popular audio + MIDI software, Pro Tools allows you to loop audio or MIDI regions or region groups using the Region menu's Loop command, which can be more flexible than using the Duplicate or Repeat commands. For example, selecting and moving a looped region selects and moves the source region and all of its repeated region loops as one.

Step 1. Select a region or region group.

Step 2. Choose Loop from the Region menu to open the Region Looping dialog.

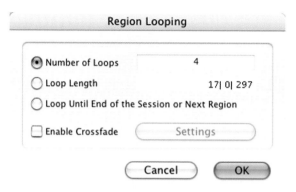

Fig. 9.49 – Region Looping dialog.

Step 3. Select the 'Number of Loops' option and enter the number of times to loop the region, enabling a crossfade at the loop point if necessary.

Alternatively, you can enter an exact duration for the loop, or loop until the next region on the track or until the end of the session.

Step 4. Click OK when you have chosen your settings and the region will be looped for the specified number of times.

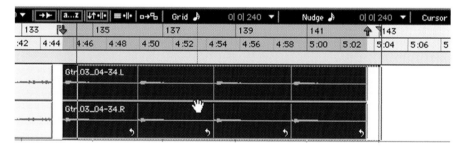

Fig. 9.50 – Looped regions with the Loop icon visible in the lower right corner.

note ▷   Looping a region does not loop any automation associated with the source region, but you can always use the 'Copy Special' and 'Paste Special' 'Repeat To Fill Selection' commands to copy automation from the source loop to all the loop repeats.

The Region menu's Unloop command lets you unloop looped regions. You can either remove all the looped regions to revert back to the original region, or you can convert the looped regions into normal regions – a process referred to as 'flattening' the loop.

Fig. 9.51 – The Unloop Regions dialog.

Editing Looped Regions

You can edit looped regions either as a group or as individual regions. To edit an individual looped region, click its Loop icon to select this first.

You can trim selections from within looped regions using the various Trim Region commands available from the Trim sub-menu in the Edit menu or you can trim the entire looped region using the standard Trim tool.

tip ▷ | If you hold down the Control key (Mac) or Start key (Windows) while using the Trim tool, it will constrain the Trim tool to removing or revealing the individual regions from the looped region.

There is also a special Loop Trimmer feature that lets you trim the duration of the individual loop regions while filling the total length of the looped region. Depending on whether you extend or reduce the length of an individual loop region, the number of repetitions of the trimmed loop increases or decreases to fill the length of the entire looped region. Let's see how this works in practice:

Step 1. Select the Standard or Scrub Trim Tool.

Step 2. Move the cursor over a Loop icon in a looped region. The cursor changes to the Loop Trimmer icon.

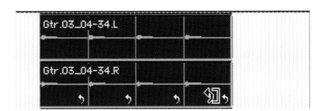

Fig. 9.52 – The Loop Trimmer about to be applied to a looped region.

Step 3. Trim the start or end of any of the individual loop regions in the group.

In the example below I trimmed the last looped region to decrease it to less than half of its previous length

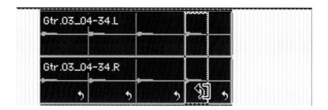

Fig. 9.53 – Using the Loop Trimmer to trim the end of the last individual looped region in the group.

Step 4. Let go of the Loop Trimmer.

As you can see in the screenshot below, all the other regions were automatically adjusted to this new length, and additional looped regions (almost 5 new looped regions) were created to fill the original length of the looped regions (making a new total of almost 9 looped regions).

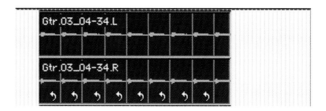

Fig. 9.54 – The looped region filled with the individual trimmed loop regions.

Step 5. Now let's go the other way and extend the length of one of the individual regions.

In the example below, it was only possible to extend the individual regions by about 60% because the audio file referenced by these regions was quite short in duration.

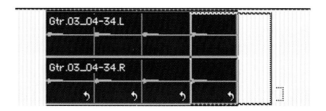

Fig. 9.55 – Extending the last looped region using the Loop Trimmer.

Step 6. Let go of the Loop Trimmer.

As you can see in the screenshot below, in this example just two whole loop regions fit within the original looped regions selection. A third, truncated, loop region is added at the end to fill up the available space.

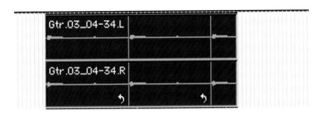

Fig. 9.56 – The looped region filled with the individual trimmed loop regions.

note ▷ If you have made a selection within a looped region, or even a selection that extends beyond the looped region, you can use the commands in the Edit menu's Trim Region sub-menu to trim the region in various interesting and useful ways. For example, if the source region is extended to the left using the Trimmer tool for the total length of the loop, the source region is moved earlier in the Timeline and repeated loop regions fill in up to the point where the last original loop region ended. If the trim to the left is part of the source region's length, the source region is not moved and a partial loop region is created to the left of the source region. This is a powerful feature that lets you make quick changes to your arrangement by using partial loops as upbeats, or by extending looped sound effects or ambience earlier in a film score.

Stripping Silence from Regions

The Strip Silence feature lets you analyze audio selections containing one or more regions across one or more tracks to define areas that you wish to regard as 'silence'. Once you have identified these, you can 'strip away' (i.e. remove) the 'silence', or you can keep (i.e. 'extract') the 'silence' by removing the rest of the audio. The third option is to separate your selection into lots of smaller regions so that these can be quantized or edited individually afterwards.

Four sliding controls in the Strip Silence window let you set the parameters by which 'silence' will be defined.

tip ▷ If you press Command (Mac) or Control (Windows) while adjusting the sliders you get finer resolution.

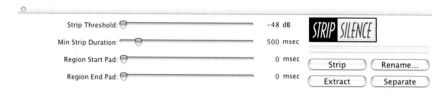

Fig. 9.57 – The Strip Silence window.

The Strip Threshold parameter lets you set a value for the amplitude, below which any audio is considered to be silence. For example, on a bass drum track, you will probably hear the sound of the rest of the drum kit when the bass drum is not playing, but this will be much quieter than the bass drum – in other words, the amplitude of this quieter audio will be much less than the amplitude of the bass drum sound. You can define this low amplitude audio to be considered as silence by setting the Strip Threshold just above this.

You can also define what is to be considered to be silence using the Minimum Strip Duration parameter. This sets the minimum amount of time that the material below the threshold must last for before it is considered to be wanted audio.

The Region Start Pad parameter lets you extend the start of each new region to include any wanted audio material that falls below the threshold, such as breathy sounds before a vocal or the sound of fingers sliding up to a note or chord on a guitar. Similarly, the Region End Pad parameter lets you extend the end of each new region to make sure that the full decay of the audio material is preserved.

As you adjust the controls, rectangles start to appear in the selected region surrounding the wanted audio. The audio in between these rectangles is considered to be silence.

For example, take a look at the screenshot below. This shows a selection from the bass drum track of a multi-mic'ed drum kit. The bass drum inside the selection plays 'bu-boom. . . .bu-boom. . . . bu-boom. . . .bu-boom'. These four sets of notes are outlined by four white rectangles. Everything in between is considered to be silence. As you can see, three additional areas of the waveform within the selection are also outlined by white rectangles. These are where the unwanted audio rises above the Strip Threshold for a short time. This was the best compromise that could be achieved by adjusting the parameters, so I accepted these settings even though it meant that I would have to manually deal with the two wrongly defined areas.

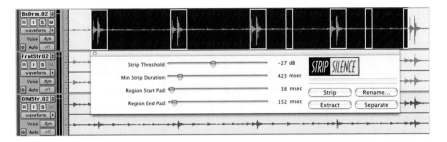

Fig. 9.58 – With the Strip Silence parameters correctly set, you see white rectangular boundaries around the 'wanted' audio.

If you click the Strip button, it removes the areas that you have defined as silence from the selected region. In the screenshot below you see the four 'wanted' regions containing the bass drum notes and the three smaller 'unwanted' regions. Now you can delete the unwanted regions and move the bass drums around to change the timing or 'feel' – or edit these whatever way you wish.

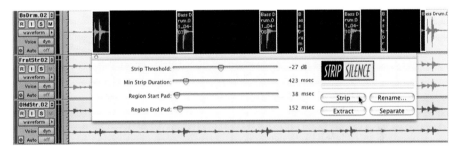

Fig. 9.59 – The 'silence between the "wanted" regions has been stripped away by clicking "Strip"'.

If you click the Extract button instead, this removes the audio above the Strip Threshold and leaves (or 'extracts') the audio that you have defined as silence. So you can think of Extract as the inverse of Strip with the 'wanted' audio in this case being the audio that falls below the Strip Threshold. This feature can be useful in post-production if you want to extract the 'room tone' or ambience from part of a recording to use elsewhere.

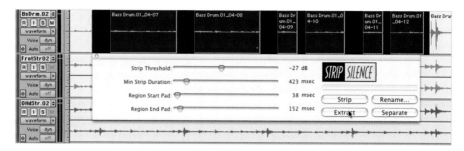

Fig. 9.60 – The audio above the Strip Threshold has been removed, leaving the audio defined as 'silence', but which actually contains ambience, remaining (i.e. 'extracted') in the track.

The third option is to click the Separate button. This automatically divides the selection into regions based on the boundaries detected by Strip Silence. In the screenshot below you see that the selection has been divided into 13 separate regions, each with its own name (although the 13th region is too short to display its name). The Separate feature is very useful if you want to quantize the audio in each region to line up with the bars and beats, for example.

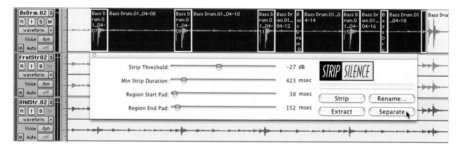

Fig. 9.61 – The Separate button divides the whole selection into separate regions based on the boundaries detected by Strip Silence.

If you click the Rename button in the Strip Silence window, this opens the Rename Selected Regions dialog. This lets you define how the regions that you will create using the Strip Silence feature will be named. For example, if you set the naming options as in the screenshot below, the names generated for the first six regions created by Strip Silence would be as follows:

- SFX023.Reel1

- SFX024.Reel1

- SFX025.Reel1

- SFX026.Reel1

- SFX027.Reel1

- SFX028.Reel1

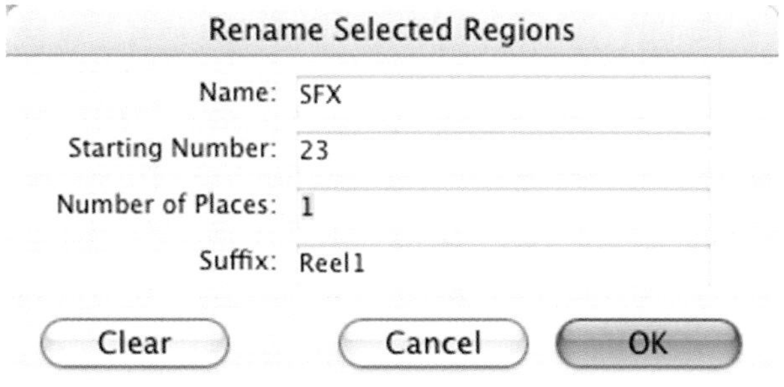

Rename Selected Regions

Name: SFX

Starting Number: 23

Number of Places: 1

Suffix: Reel1

Clear   Cancel   OK

Fig. 9.62 – Rename Selected Regions dialog.

note ▷   This dialog remembers your previous settings. You can clear any unwanted previous settings using the Clear button.

Inserting Silence

The Insert Silence command does what it says, replacing a selection that you have made on a track or tracks with silence. Less obviously, it can also be used to remove automation data.

If the track is displaying audio or MIDI data, when you apply the Insert Silence command to a selected range, it not only clears the audio or MIDI data, it also clears any automation data for the track or tracks.

However, if the selected tracks are displaying automation data, the automation data visible on each track is cleared throughout the selected range and any audio or MIDI is left untouched. Also, if you press the Control key (Mac) or Start key (Windows) while choosing the Insert Silence command, this clears all the automation data from all the selected tracks – not just the visible data.

The Insert Silence command is particularly clever when used in Shuffle mode. In this case it moves the track data within and after the Edit selection forward, by an amount equal to the selection, pushing everything (including any automation data) forward from the start of the selection – to get it 'out of the way' of, thus making room for, the silence that you are inserting.

To explain this another way: when you apply the Insert Silence command to an Edit selection while in Shuffle mode, Pro Tools splits the region(s) at the beginning of the insertion point and moves the new region(s) to a position later in the track by an amount equal to the length of the selection, before inserting the selected amount of silence.

tip ▷   If this written description still leaves you wondering what this is all about, then the best way to get to know how this works (as with most things) is to try it for yourself.

Consolidating Regions

Often, during your editing sessions, you will end up with tracks made up from many smaller regions. This could happen if you 'comp' several takes to make a composite containing the best from each take. Or maybe you have separated several small regions from one take containing notes unintentionally played 'off the beat' so that you can move these to their correct bar/beat positions.

When you are satisfied with your edits to these tracks, Pro Tools allows you to consolidate a track, or a range within a track (such as a verse or chorus), into a single region, which is much easier to work with. When working with audio tracks, this consolidation process causes a new audio file to be created that encompasses the selection range, including any blank space, treating any muted regions as silence. To consolidate regions within a track:

Step 1. Select the regions you want to consolidate using the Time Grabber tool or the Selector tool.

Step 2. Choose Consolidate from the Edit menu to create a new, single region that replaces the previously selected regions.

tip ▷ Consolidating an audio track does not apply any automation data as it creates the new file, so if you want to create a new file with automation data applied to the audio, use the 'Bounce to Disk' feature instead.

Compacting Audio Files

When you have finished your editing session you should consider 'compacting' any edited audio files to remove any unused audio that you are sure you have no further use for. This makes the files smaller, so it saves disk space from being used up unnecessarily.

If you have recorded or imported a large number of audio files, yet have only actually used a fraction of these, compacting the files can save considerable amounts of disk space, making backups much less costly both in time needed to make the backups and for the media used.

To compact an audio file:

Step 1. Select the region or regions that you want to compact in the Region List.

Step 2. Choose Compact from the popup menu at the top of the Region List.

Step 3. Enter the amount of padding in milliseconds that you want to leave around each region in the file.

Step 4. When you click Compact, the file or files are compacted and the session is then automatically saved.

note ▶ Because it permanently deletes audio data, the Compact Selected command should only be used after you have completely finished your editing and are sure that you have no further use for the unused audio data.

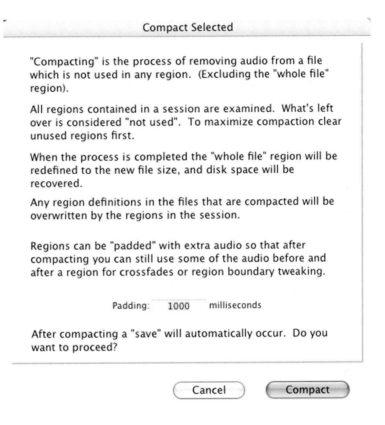

Fig. 9.63 – The Compact Selected dialog.

What you have learned

How to use the Cut, Copy, Paste, and Clear commands and how to make Edit selections are explained in some detail.

The various Edit modes are explained along with tips and tutorials on how to use these.

Making Timeline selections and the pros and cons of linking these with Edit selections are explained.

How to find your way around the Edit window, zooming, scrolling and finding the ends of the selection are explained in some detail.

How to work with regions and how to use the various Region Editing commands.

Finally, how and when to consolidate and compact regions and audio files is explained.

Mixing

Overview

Mixing is an art – not a science. There are no rules to this game that cannot – and have not been – broken. Nevertheless, there are various aspects that most mix engineers generally agree on.

The single most important activity is balancing the levels of the individual instruments and voices against one another. I cannot stress how crucial this balancing is. A recording can be completely transformed when you adjust or re-adjust the balance between the instruments – for better or for worse.

If you are mixing a song, the lead vocal is usually going to be the most important element in the mix. Similarly, if it is an instrumental, the featured lead instrument – trumpet, guitar, saxophone or whatever – becomes the most important element. You must make sure that the lead stands out, stays at a consistent and audible level throughout, and sounds great, using a combination of EQ, compression, reverb, delays, and other effects as appropriate. Some engineers like to start by working on the lead vocal or instrument first, before bringing in the rest of the instruments.

In many forms of popular music, the rhythm section is almost as important, and in dance music it is sometimes as important or even more important than the lead. If this is the case, you will probably want to mix the rhythm section first before bringing in the vocals.

Once you have a rough balance with the vocal 'sitting' nicely on top of the rhythm section, you might want to add in various other elements: backing vocals, percussion, brass, strings, synthesizers, or other instruments.

You might also want to consider leaving some of the mix elements out in the earlier parts of the song so that the arrangement 'builds' in a more interesting way than if you hear everything straight away. You can mute these elements then unmute them where they are needed.

Panning the various instruments to different positions within the stereo space also helps to develop a more interesting mix. For example, it is usual to place the bass drum and bass guitar dead centre in the mix – although this was not the case during the first few years of stereo mixing when the entire rhythm section might be panned hard left with brass and strings panned hard right and the vocals occupying the whole of the central space. Arguments rage between producers as to whether the drum kit should be panned so that the listener hears it as the drummer would or as a member of the audience would. Strings, woodwinds, and brass are often panned as in the standard orchestral layout as heard from the audience with high strings left, low strings right, horns left, brass right, and woodwinds centre. But, again, there are no rules – it is a question of the preference of the producer.

You may want to use three or more reverb units, plus delay units, and harmonizers to help to blend the instruments together and to help to create the illusion that separately overdubbed musicians are all playing together on the recording. You can use harmonizers on backing vocals, on electric piano 'pads', on guitar solos – anywhere that you want to add depth to your mix. A little slapback echo always seems to help a vocal to stand out in the mix, and judicious use of different reverb types helps to build a more interesting 'sound picture' for your mix.

There was a time when several pairs of hands were needed on the mixing console to carry out a complex series of mix manoeuvres in real time before the advent of mix automation. Automated mixing makes life much easier. By recording the fader movements, you can set up your mix while concentrating on just a couple of faders, then on the next pass, work on a different pair of faders, and so on until you have your mix balanced the way you want it. And it is just as easy to automate the pans and mutes, vary the EQ, shorten or lengthen the reverb decay, or apply changes to other effects during the course of the mix.

Finally, you may want to create a long, smooth fadeout at the end of the song, ideally using a physical fader to control the Master Fader in Pro Tools. You can do this using the mouse onscreen, and you can certainly achieve excellent results by drawing in the automation curve by hand. But nothing beats the feel of a high-quality fader under your finger as you feel out your final fade. . .

Monitoring

One of the most important things to attend to is your monitoring system – you have to be able to hear what you are doing properly. This means making sure that you are sitting in the 'sweet spot' at one of the three corners of an equilateral triangle with the left and right speakers positioned at the other corners facing you, and with the tweeters at about the same height as your ears.

You should make sure that the amplifiers have more than enough power to reproduce sound at the highest listening levels you intend to use without significant distortion and that the speakers have as 'flat' a response as possible over as wide a range as possible. So-called 'nearfield' monitors (such as the ubiquitous Yamaha NS10's) that sit on stands, on desktops, or on the mixing console's meter bridge, will never be full-range speakers – because they have to be relatively small. This means that they will not be able to reproduce the bottom octave properly as their response will roll off significantly below, say, 100 Hz. This is why mastering studios, for example, use large, full-range monitors to make sure that the mastering engineer can hear all the frequencies correctly.

It is always a good idea to listen to your mixes on the kinds of speakers in the kinds of environments that will ultimately be used by your audience. So you might take your mixes out to a club, or play them in your car, or try them on your living room system – you get the idea. And don't forget to check your mixes in mono – especially if you want them to sound right on radio, TV, or film where mono playback is still sometimes encountered.

Monitoring Levels

There are no standard monitoring levels observed in music recording studios – unlike in film dubbing theatres, where the Society of Motion Picture and Television Engineers (SMPTE) has established a Sound Pressure Level (SPL) of 85 dB as a standard. This is referenced to an electrical signal level of –18 dBFS, which is the standard Operating Level recommended for digital systems by the Audio Engineering Society (AES). I have been at sessions in music studios that use monitoring levels well up into the 90's of decibels of SPL – or even more than 100 dB. Even relatively short exposure times to such high levels can make your ears 'sing' due to temporary threshold shift. Nevertheless, you may wish to check out the sound of some loud instruments that you have recorded at somewhere near to their original sound levels. And instruments such as the trumpet or snare drum can easily reach levels of over 100 dB, so it can be useful to have a monitoring system capable of reproducing such realistic levels.

Many engineers like to mix on small 'nearfield' monitors at quite low levels at somewhere between, say 65 and 75 dB SPL. This makes it easier to tell whether the lead vocal or lead instrument can be heard properly at all times and whether other important elements such as the bass guitar and snare drum are at the right levels in relation to the lead. It also means that the mix should sound great when played back on domestic speakers after the record is released to the public. Of course, you can always check how it sounds from time-to-time at much louder volumes or on larger speakers.

Metering

Each Pro Tools track has its own metering running vertically next to the track's fader. These meters let you see at a glance what the individual signal levels are and what the summed signal levels are at the Master Faders. On audio tracks, the meters show the levels of the signal being recorded or played back from the hard drive. On Auxiliary Inputs, Instrument tracks, and Master Faders, the meters indicate the level of the signal being played through the channel output(s). Signal present is indicated by a green colour that turns to yellow when the level reaches 6 dB or less below full scale. Each meter also has a Clip Indicator at the top that lights up red if clipping occurs.

These meters show levels in dBs relative to the nominal (zero) operating level. So what's all this about? Let's look at some of the concepts involved. The meters used in Pro Tools systems are similar to the 'VU' or 'Volume Units' meters used in analogue audio systems in that they represent an average measure of the signal levels that corresponds reasonably well with the way the human ear perceives the volume of the reproduced audio. They do not reveal the peak levels of the audio signals, which may be 12 or more dBs higher than the average levels. Because these peak levels are typically very short in duration they are often referred to as 'transients', meaning that they pass quickly.

note ▷
> The nominal Operating Level is the 0 VU level that is used as the standard input and output level for signals entering and leaving the system. This level may be set differently in different audio systems, and is typically set some way below the maximum level that the system can handle.

In practice, signals passing through a system may be lower or higher than the nominal operating level. If signals are much lower than 0 VU, these transients may get too close to the background noise floor that exists in most systems – which may mask some elements of the wanted sound. On the other hand, if signals passing through the system get much higher than 0 VU the transients may exceed the maximum allowable level and the sound will become distorted.

To allow for the transients to pass undistorted while using VU style metering, audio systems use the concept of 'headroom'. In analogue systems, for example, the 0VU level is set such that signal peaks may reach 12 dB (or more) above 0 VU before distortion occurs. And a characteristic of analogue systems such as tape recorders is that the distortion is mild at first, so even higher levels can be used before obnoxious distortion is heard.

In the case of digital systems, including Pro Tools, the Audio Engineering Society (AES) standard is to set the nominal (normal) Operating Level at 18 dB below the '0 dB Full Scale' or '0dBFS' level which represents the maximum allowable signal level in the system. If any signals exceed this 0dBFS level they will be 'clipped' to this level on input to the system, distorting the signal extremely unpleasantly. Notice that the headroom available in these digital systems is 18 dB, which is much greater than with most analogue systems.

note ▷
> The difference between the nominal Operating Level and the level of background noise inherent in the system (often called the 'noise floor') is referred to as the 'signal-to-noise ratio'. Typically, digital systems have a much lower noise floor than analogue systems. As a consequence, you are more likely to hear quiet sounds such as rattles or squeaks from musical instruments or drum kits, or breathing or other sounds made by musicians, that you may not have noticed with analogue recordings where these low-level sounds may have been masked by the general background noise, tape hiss, and so forth.

If you connect your Pro Tools system to any analogue devices, such as mixers, tape recorders, or effects devices, you must make sure that the nominal Operating Level of your Pro Tools system matches the nominal Operating Levels of these devices. In other words, if the meters in Pro Tools read 0 dB the meters in any connected equipment should read 0 VU.

To check that connected equipment is lined up correctly, you can insert the Real-Time AudioSuite (RTAS) Signal Generator plug-in into one of the Master Fader Inserts, generate a test 'tone' such as a 1 kHz sine wave at 0 dB, and see whether the meters on connected equipment match this.

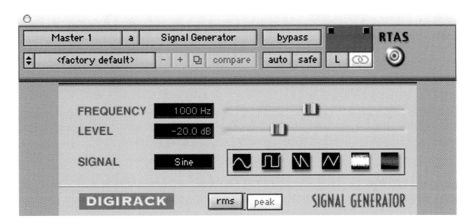

Fig. 10.1 – Using the Signal Generator plug-in to generate a test tone.

Alternatively, you can use the Signal Generator AudioSuite plug-in to generate a 1 kHz sine wave at any suitable level, for example –20 dB. You can either record this 'tone' just before you mix onto tape or into a separate digital file that you supply along with your mixes. This can then be used as a reference tone that other equipment can be aligned to in order to establish the correct value for 0VU.

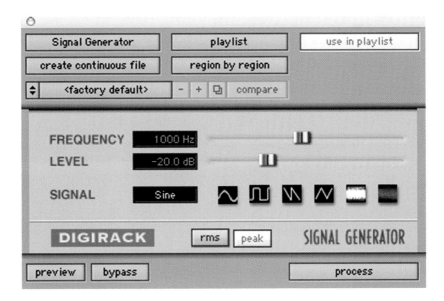

Fig. 10.2 – Using the Signal Generator AudioSuite Plug-in to generate a test tone.

If you are connecting your Pro Tools system digitally to external devices such as mixers, DAT recorders, effects units, and so forth, as long as these use the same digital standards, then signal levels sent from Pro Tools will produce the same meter readings on the destination devices and vice versa.

note ▷ Theoretically, the dynamic range of a 16-bit digital system is 96 dB while that of a 24-bit system is 144 dB. The dynamic range is the range between the minimum and maximum signal levels that the system can handle.

Fat Meters

When you are mixing you may prefer to use the 'fat' meters instead of the normal 'thin' meters in the Mix (and Edit) windows.

tip ▷ A useful keyboard command lets you easily switch between normal and 'fat' meters: hold the Command (Start), Option (Alt) and Control keys and click on any of the meters.

Master Fader Tracks

Although it is possible to create a Pro Tools Session that does not have any Master Faders, that just has audio tracks routed directly to audio interface outputs, it is normal practice to have at least one stereo Master Fader track to control the audio before it is routed to the audio interface's main stereo outputs. This Master Fader can then be used to fade out the level at the end of a mix pass.

Sessions may have additional Master Fader tracks to control audio feeds to headphone circuits for the musicians, or to control effects send levels to external effects units, or for other similar purposes.

To set up a Master Fader as a stereo master level control:

Step 1. Choose 'New. . .' from the Track menu to open the 'New Tracks' dialog and create one new stereo Master Fader.

Fig. 10.3 – Creating a new Master Fader.

Step 2. Set the outputs of all the audio tracks in the session to the main output path (e.g. outputs 1–2 of your primary audio interface) and set the panning for each track.

Step 3. Set the output of the Master Fader to the main output path.

Master Faders allow you to use up to five post-fader inserts. You would typically use these to insert EQ, compressor or limiter plug-ins, or a dither plug-in on your master mix.

Multiple Output Assignments

As an alternative to using multiple Master Faders to output audio to different physical output paths, it is possible to directly assign Pro Tools audio, Auxiliary Input, or Instrument tracks directly to multiple output paths.

Assigning track outputs to multiple paths lets you efficiently route an identical mix to simultaneous monitor feeds, headphone mixes, or wherever this may be needed.

To add an additional output assignment to an existing track output assignment, Control-click (Mac) or Start-click (Windows) as you select the additional output path in the track's Output Path selector. A plus sign (+) is added to the name of the assigned output to indicate that the track has multiple output assignments.

Fig. 10.5 – Multiple track output assignment indicated by the plus (+) sign.

Fig. 10.4 – Master Fader with 4-Band EQ, compressor, limiter, POWr dither, and Signal Generator Inserts.

tip ▷ If you hold the Option key (Mac) or Alt key (Windows) as well, the additional output assignment will be added to all tracks. And if you also hold the Shift key, the additional output assignment will be added to all selected tracks.

Making Tracks Inactive

When you start mixing your tracks in Pro Tools, you need to 'husband' your DSP resources very carefully with all but the simplest of sessions. For example, if the session includes one or more tracks that may not be included in the mix, it is best to make these tracks inactive until and unless you decide to use them.

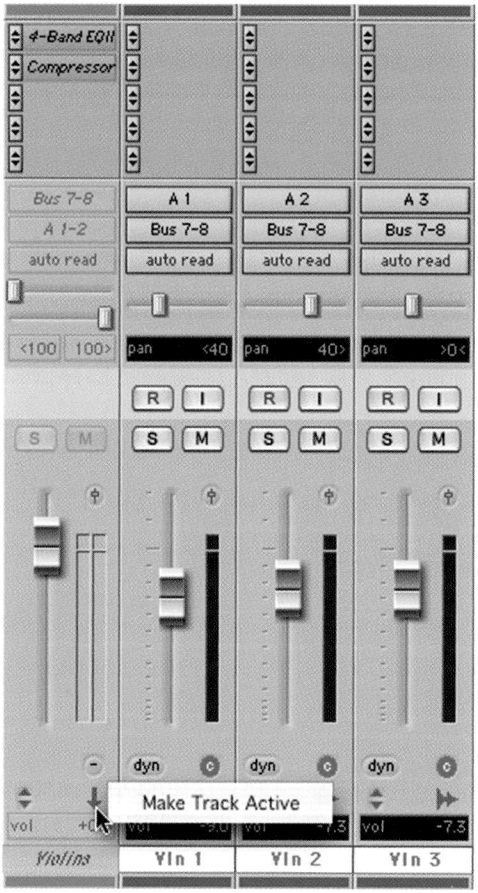

It is possible to set Track outputs to 'No Output', but this does not free the associated DSP resources. Inactive outputs do not consume resources for mixer connections, but any assigned plug-ins on the track continue to use their required DSP resources. You can, of course, free up the DSP used by the plug-ins by making the plug-ins inactive.

A simpler way to deal with the situation is to make the whole track inactive – which frees up the DSP used for both mixing and for the plug-ins. Each track has an icon to identify the track type located just below the track meter. This icon also serves as a popup selector that you can use to make the track inactive (and vice versa). Inactive tracks are 'greyed out' so that you can see at a glance that they are inactive.

Fig. 10.6 – Make an inactive track active (or vice versa) by selecting from the popup selector underneath the track meters.

In the example shown in the accompanying screenshot, with three violin tracks submixed into an Auxiliary track, it would also make sense to make the individual violin tracks inactive as well as the Auxiliary track if the violins are not going to be used in the mix.

Grouping Tracks

At the start of your mixing session, if you have not already done so, it makes sense to group particular sets of tracks together. For example, you might have a set of several drum tracks, or a set of brass instruments or strings, or maybe a set of three or four guitar parts. When you have decided on the relative balances between the individual members of these sets you will almost certainly want to adjust the overall balances of these sets of instruments within your mix.

Grouping the members of the sets together lets you move one fader from the group and have the other group members follow exactly. And if you need to make selections and use cut and paste or make other edits, these will apply to all members of the grouped set of instruments. For example, if you press the Solo or Mute buttons for any member of the Group, the whole Group will become soloed or muted.

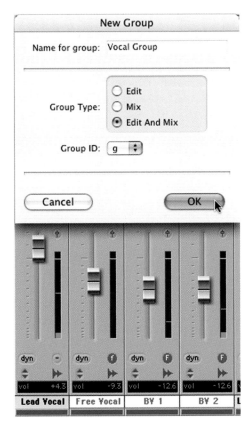

Fig. 10.7 – Grouping Tracks.

To create a Group:

Step 1. Select the tracks of interest, Shift-clicking to add to your selection without losing your previous selection.

Step 2. Choose 'Group. . .' from the Track menu or Press Command-G (Mac) or Control-G (Windows) to open the New Group dialog.

Step 3. Name the group, select a Group ID, and choose whether to apply the group to both Edit and Mix windows or just one of these.

Step 4. Click OK to create the new group.

Pro Tools has 26 memory locations that can be used to store Groups, each of which is identified by a letter of the alphabet. You can turn Groups on or off simply by pressing the appropriate letter on your computer keyboard: with the Keyboard Focus feature enabled, simply type the Group ID letter (any of the lowercase keys, a–z) on your computer's keyboard. The Keyboard Focus is always enabled in the Mix window, but in the Edit window it is disabled by default, so you need to specifically enable the Key Focus if you want to use this feature. To enable the Edit Groups List Keyboard Focus you just click on the small 'a–z' button in the upper right corner of the Edit Groups List.

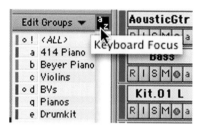

Fig. 10.8 – Edit Groups List just after clicking on the 'a–z' button to enable the Keyboard Focus.

tip ▶ If you need to adjust one fader within a Group, simply hold the Control key (Mac) or Start key (Windows) on your computer keyboard and you can tweak this on its own. When you let go of the key, you have the Group back in operation again.

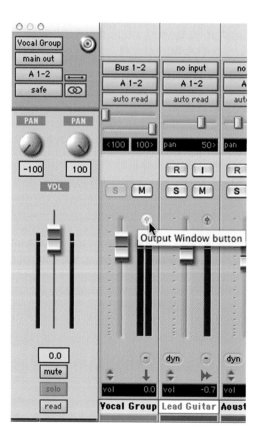

Track Output Windows

Pro Tools provides dedicated Output windows for track outputs. To open a track Output window you simply click the Output Window button which you will find a little way above the track meter to the right of the fader on each channel strip. The track Output window provides an alternative, clearer display for setting the various track output parameters: the level faders and rotary pan controls, mute, solo, and Automation mode buttons.

Fig. 10.9 – Here you see the Vocal Group track's Track Output Window at the left with the Mix window visible to the right of this. The cursor is about to click on the Output Window button (which will close/open the Track Output Window).

note ▷ Track Output windows have a red 'target' button that you can use to keep the window open while you open other output windows. Click on this and it will turn from red to grey – and this window will stay open when you open another window. To keep each new window open you need to click in its target button as well. Having lots of these Track Output windows open uses up a lot of screen space, but can be useful at times.

Using Auxiliary Tracks

Although Auxiliary tracks cannot playback audio from disk, they can be used for a variety of useful purposes in Pro Tools. The most basic use is, perhaps, to feed inputs from external sources such as synthesizers or effects units connected to your hardware interface into your mix.

A more typical use for Auxiliary tracks is to accept output signals via internal busses from a group of tracks to form a submix that you can control using one fader. You can then apply effects and processing to this submix using just the one set of Inserts on the Auxiliary track instead of applying these individually to each track within the group. So, for example, you might submix the drums, the vocals, or any other logical groupings of instruments.

Pro Tools lets you insert up to five inserts on each Auxiliary Input track. Each insert can be either a software plug-in or can be an external hardware device hooked up to the insert via your Pro Tools hardware interface. Some Instrument plug-ins (such as Virus Indigo) also accept audio from the track input, so you can also use these as signal processing plug-ins. Both plug-ins and hardware inserts route the signal from the track through the effect and automatically return it to the same track.

To create a submix using an Auxiliary track:

Step 1. Create a new stereo Auxiliary track.

Step 2. Select an available bus pair as the input to this Auxiliary track.

Step 3. Route the outputs of the audio tracks that you want to submix to this same bus pair so that they form the input to the Auxiliary track.

Step 4. Balance the levels and set the pans to create the submix.

Step 5. Insert and apply plug-in or external hardware effects to the Auxiliary track as desired.

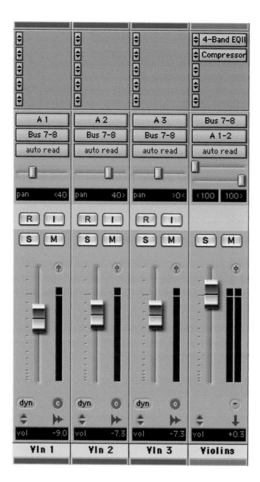

Fig. 10.10 – Violins submixed to an Auxiliary track via bus paths 7 and 8 with EQ and compression plug-ins applied to the submix via the Auxiliary track's inserts.

In this arrangement, the balance between the processed and unprocessed signals is controlled by the plug-in's (or by the external hardware's) wet/dry controls. The contributing track faders control the balance within the submix while the Auxiliary Input track fader controls the output levels of all the tracks routed to it.

tip ▷ You can always bounce a submix to disk to free up the voices from the contributing tracks for use by other tracks.

note ▷ Using track outputs to assign tracks to bus paths to feed an Auxiliary track is the preferred submixing configuration when you want to use dither or other mastering processing, in which case you do not want unprocessed audio to be heard at the same time as the processed signals. If you used Sends to route signals to the Auxiliary track instead of using the track outputs to route signals to the Auxiliary track, the unprocessed audio would still be routed via the track outputs to the main stereo mix outputs so you would hear this as well as the processed audio from the Auxiliary track.

Using Sends

Another way to route signals in the Pro Tools mixer is by using the Sends. As their name implies, these let you send signals from a track to a destination of your choice. Sends are to be found on most professional mixers and the way they are implemented in Pro Tools is particularly flexible.

You can insert up to 10 sends (labeled A–E and F–J) on each audio track, Auxiliary Input, or Instrument track. These can be set as pre- or post-fader and the send levels, mute status (and send pan status with stereo sends) are fully automatable.

Sends are displayed in the Mix and Edit windows according to the options you select in the View menu for Sends A–E and Sends F–J. If you are using five or fewer sends, you only need to display and use Sends A–E. If you need more than five Send Assignments, then you can use Sends F–J as well. You can choose to display all five Assignments in each group to give you an overview, or you can display the level, pan, mute and pre/post-fade buttons and send level meter for any of the individual sends to see more detail.

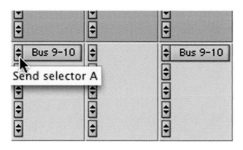

Fig. 10.11 – Default Assignments View in the Sends section of the Mix window showing Send A assigned to bus pair 9–10.

The default view shows the five Assignments. You can select individual Send Views from the View menu or you can Command-click (Mac) or Control-click (Windows) on the Send Selector to switch the display (on all tracks) to show the details of that individual Send.

Individual Send Views (such as Send A) display all the controls of an individual send in the sends area of both the Mix and Edit windows, providing full access to all controls for that send on all tracks.

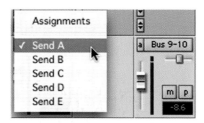

Fig. 10.12 – Individual Send A displayed in the Sends section of the Mix window.

tip ▷

If you want to switch the display back to showing the Assignments view when you are displaying sends in an individual Send View (such as the Send A View), just Command-click (Mac) or Control-click (Windows) on the Send selector and select 'Assignments' from the popup that appears.

The Send level and mute controls can be set to follow Mix groups, allowing you to adjust multiple send controls from a single set of controls. To set this up, you need to open the Automation Preferences window and tick the options for 'Send Mutes Follow Groups' and 'Send Levels Follow Groups'. With these options selected, if you adjust the Send Level or change the Mute status on any track within a group, the Send Level and Mute status for all tracks within that group will follow these changes.

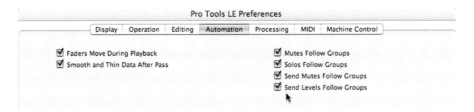

Fig. 10.13 – Selecting 'Send Mutes Follow Groups' and 'Send Levels Follow Groups' in the Automation Preferences window.

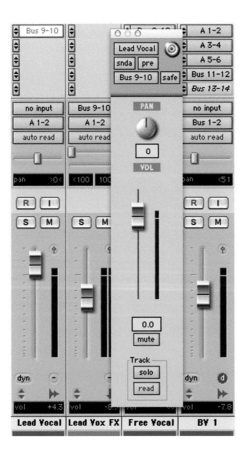

You can also open individual Output windows for each send. These are similar to the track Output windows. These output windows present all the Send controls in a convenient strip so that you can set the send level, pan, and mute controls, solo the track or change the track automation status, 'safe' the send to prevent it from being accidentally automated, and select its pre/post-fader status.

Fig. 10.14 – Pro Tools Mixer showing Sends A–E in Assignments View with the Output window open for the Lead Vocal Send.

Sends can be assigned to any available output or bus paths in mono or stereo by clicking on and holding the Audio Output Path selector.

Fig. 10.15 – Audio Output Path selector.

The popup Audio Output Path selector then lets you assign the Send to any of the outputs on your hardware interface or to any of the internal bus paths.

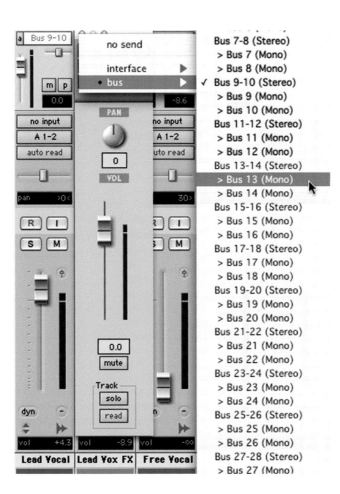

Fig. 10.16 – Assigning a send to a mono bus path.

tip ▷ Why not label your buses using the I/O Setup window? If you know you will always be using a particular bus pair as inputs to an Aux channel with a reverb inserted across it, you can label this 'Reverb' or some other suitably descriptive name. This will make it easy to select the correct bus when you are routing tracks to this reverb.

Automating Sends

To enable you to control the levels and placement of effects with the greatest precision during your mix sessions, Pro Tools lets you automate the level, mute, and pan settings for the sends.

Step 1. Check that the send level and/or send mute and/or send pan is enabled in the Automation Enable window.

Step 2. Make sure that the Automation mode for each track with sends that you want to automate is set to Write for your first automation pass.

Step 3. Make sure that the sends are currently visible on your tracks using the View menu to select Sends A–E (and Sends F–J if you are using these).

Step 4. Open the Output window for the send you want to automate by clicking on the send, or display the individual view for the send in the Mix or Edit window.

Step 5. Play the session and move the controls you want to automate.

Step 6. Click Stop when you have finished.

Returns

If you send signals from a track to some destination, you will usually want to return the signals to your mix from that destination. You can return signals routed using Sends back via Auxiliary Inputs or Instrument tracks (or Audio tracks) into your mix.

note ▷ Signals routed via Sends are usually monitored and processed using an Auxiliary Input or Instrument track because you will lose one of your available 'voices' if you use an Audio track for this purpose.

If the Sends are routed to an audio track, the audio being sent can be recorded onto that audio track. However, any audio signals being routed via busses can always be bounced to disk by choosing these busses as the bounce source in the 'Bounce To Disk' dialog. So, if you want to record the audio returning from these Sends to your hard disk so that you can use the newly created audio files elsewhere in your session, or even in another project, it is more efficient to use the Bounce To Disk feature instead.

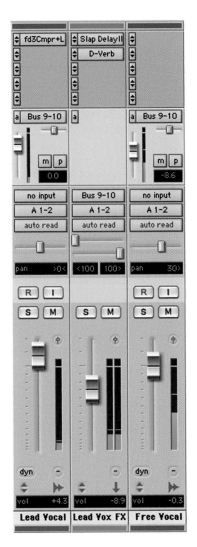

Send and Return Submixing for Effects Processing

The most typical use of Send and Return submixing via the internal busses is to let you add reverb, delay, or other similar effects to a group of tracks (or perhaps even to all of the tracks).

Send and Return bussing lets you use a real-time plug-in or a hardware insert as a shared resource for all the tracks included in the submix.

In the example shown, the Lead Vocal and Free Vocal tracks are routed to the Lead Vox FX Auxiliary track via Send A along bus pair 9–10. The audio sent to the Auxiliary track is then processed using the Slap Delay and D-Verb plug-ins.

Fig. 10.17 – Lead Vocal and Free Vocal tracks routed to the Lead Vox FX Auxiliary track via Send A along bus pair 9–10 for processing using Slap Delay and D-Verb plug-ins.

In this configuration, the wet/dry balance in the mix is controlled using the Lead Vocal and Free Vocal track faders to set the dry level and using the Lead Vox FX Auxiliary Input fader to set the wet (effect return) level. Note that the plug-in (or hardware) effect is normally set to 100% wet in this configuration.

When the Send is set to pre-fader (the default), the amount of effect is controlled both by the level of the Send Level fader and by the level of the Return fader – which, in this case, is the Auxiliary Input's main volume fader.

If the Send is set to post-fader, the audio track's main fader controls how much audio reaches the Send before the audio is routed via the Send Level fader along the bus to the Auxiliary track. As with pre-fade Sends, the amount of effect that you hear is also controlled by the level of the Return fader (the Auxiliary Input's main volume fader).

tip ▷ Often you will want to solo tracks when setting up your Auxiliary channels with effects to process these tracks. It can be inconvenient to have to remember to solo the Auxiliary track as well as the individual track or tracks that are being bussed to that Aux channel. A 'solo safe' feature lets you 'warn' the Aux channel to switch into solo as soon as any of the tracks being bussed to this are switched into solo. To enable this solo safe mode, simply Command-click (Mac) or Control-click (Windows) on the Solo button on the Auxiliary channel and this will turn a darker shade of grey. Now when you solo any track being bussed to this, you will hear the output of the Aux channel as well – without having to click on its solo button. I usually leave my Aux channels in this mode throughout my mixing sessions.

Automation

You can automate virtually every function in Pro Tools. Just choose an automation write mode from the popup and play the track. When you move any faders or other controls, your 'moves' will be recorded. It is as simple as that.

Specifically, Pro Tools lets you automate volume, pan, and mute controls for audio, Auxiliary Input, and Instrument tracks, along with Sends and real-time plug-in controls. Master Fader tracks only support volume and plug-in automation. MIDI and Instrument tracks support MIDI volume, pan, and mute automation.

When you stop playback, you can view and edit these automation 'moves' graphically in the Edit window. It is very easy to make or edit your automation 'moves' in the Edit window by setting control points or 'breakpoints' manually on the automation line using the Grabber tool – positioning these to achieve the effects you want. And you can make the changes extremely accurately if you zoom in far enough.

When you first create a track it defaults to Auto Read mode and puts a single automation breakpoint at the beginning of each automation playlist display. You can move a fader (or any other automation control) and the initial breakpoint will move to this new value and stay there until you move the fader again.

tip ▷ If you want to permanently store the initial position of the fader or other control you can manually place a second breakpoint after the initial breakpoint at the value you want. Alternatively, you can simply put the track in Auto Write mode and press Play – then press Stop a short time later. If you look at the automation playlist display you will see that this action has inserted a second breakpoint at the time you pressed Stop. Now, if you inadvertently move the control later, it will always return to this value when in Auto Read mode.

The way Pro Tools handles mutes and plug-in bypasses is very neat. Hold the Mute or Bypass button while in any automation writing mode and the Mute or

Bypass will be enabled for as long as you hold this down. If you want to clear any of these, just do another automation pass and hold the button again where you want to clear it. To make this even easier, the Mute button, for example, will become highlighted whenever you are passing though a muted section so when you see this you will know where your muted sections are.

The best way to fully understand how the automation works is to write some and look at the automation playlist display to see what breakpoints have been inserted. How do you do this? It's easy. In the Edit window, use the popup Track View selector in the middle of the controls at the left of each track to switch from the waveform display to display the automation type you are interested in – volume, pan, or whatever. Here you will see a line with breakpoints where you have written automation. The first thing you will probably want to do is to delete any breakpoints that are obviously in the wrong positions. You can drag the breakpoints to new positions, but if the new position is not close by, it is quicker to delete one breakpoint and insert a new one using the Grabber tool. This turns the cursor into a pointing finger when you are in one of these displays. Click on the automation graph line and a new breakpoint will be inserted. To remove a breakpoint you can Option-click (Mac) or Alt-click (Windows) on it.

tip ▷ If you want to remove several breakpoints it is quicker to change to the Selector tool and drag across the range of breakpoints you want to remove from the graph line to select these – then hit the Delete or Backspace key.

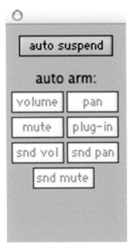

Creating Automation

Let's run through the basic steps to create automation in real time.

Step 1. First you need to make sure that the automation type that you want to record – volume, pan, mute, plug-in automation, send volume, send pan, or send mute – is enabled. Open the Automation Enable window where you will see buttons for each of these. When they are highlighted, the automation type is enabled.

Fig. 10.18 – Automation Enable window.

note ▷ To suspend writing of all automation on all tracks, click the Auto Suspend button. To suspend writing of a specific type of automation on all tracks, click the button for that automation type.

Step 2. If you want to automate a plug-in, you will also need to enable the individual plug-in parameters that you want to automate. Open the plug-in window and you will see a button marked Auto.

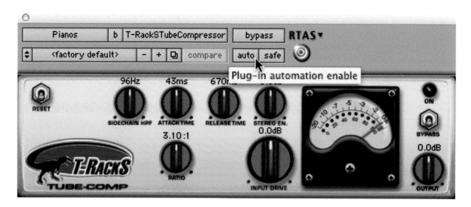

Fig. 10.19 – A DigiRack plug-in window showing cursor about to click on the Auto button.

Step 3. Click on the plug-in's Auto button to bring up a dialog box where you can choose from the list of automatable parameters for that plug-in.

Step 4. Select a parameter such as Bypass on the left and click 'Add' to add it to the list of parameters to be automated on the right.

Step 5. OK the dialog to make your chosen parameters active.

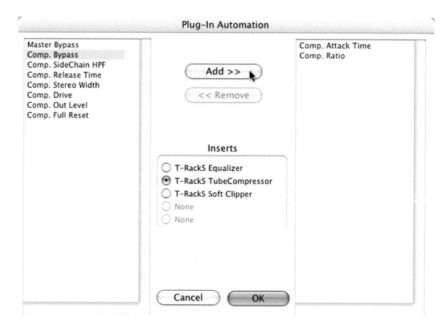

Fig. 10.20 – Plug-in Automation window.

tip ▷ As an alternative to using the Plug-In Automation window, you can enable individual plug-in controls directly from the plug-in's window by Command-Option-Control-clicking (Mac) or Control-Alt-Start-clicking (Windows) on the control. There is also a shortcut to enable all parameters for a single plug-in in one step. First make sure that the plug-in window is open for the plug-in you want to automate. Then Command-Option-Control-click (Mac) or Control-Alt-Start-click (Windows) on the Auto button at the top of the plug-in window.

Step 6. The next step is to put the appropriate tracks in an automation writing mode – choosing Auto Write, Touch, or Latch mode.

Fig. 10.21 – Selecting an Automation mode.

note ▷ An Automation Safe button is provided in the plug-ins, Track Output and Send Output windows. You can enable this if you want to protect any existing automation data from being overwritten – a wise move to make once you are happy with the way your automation is working.

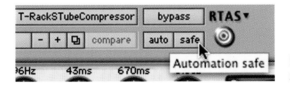

Fig. 10.22 – The Automation Safe button.

Step 7. Once you have everything set up correctly, just hit Play and move your controls as you like – then hit Stop. When you next play back, Pro Tools remembers all the moves you have made on enabled controls. It's as simple as that.

Step 8. Normally you will choose Auto Write mode for your first pass and use Auto Touch or Auto Latch modes to make further adjustments. Start playing back from wherever you like and simply move the control where you want to make your changes. New data will only be written when you actually move the control – the original data will not be altered anywhere else.

To edit automation once it has been recorded, you can rewrite new data over the previous data, or edit the automation graphically or using the cut, copy, paste, or delete automation data commands.

Automation Write Modes

Auto Write does what it says – it lets you write automation data when you move any control until you stop playback – erasing any previous data up to this point.

Auto Write mode automatically switches to Auto Touch mode when you stop, ready for a second pass to fine-tune your automation pass. In Auto Touch mode, automation is only written when you actually operate any of the controls – and the control will return to its previously automated position when you let go, at a rate that you can set using the AutoMatch Time and Touch Timeout settings in the Automation Preferences window.

note ▷ The Automation Preferences window also has an 'After Write Pass, Switch To' preference that lets you choose which Automation mode Pro Tools will automatically switch to after an automation pass in Write mode. This lets you choose to switch to Touch or Latch mode, or to stay in Write mode by selecting No Change.

In Touch mode, touch-sensitive fader control surfaces such as the Digi 002, Command|8, or the Mackie HUI start writing automation as soon as you touch them. With other control surfaces in Touch mode, writing of automation does not begin until the fader hits the pass-through point, or the previously automated position. Once you reach the pass-through point with the fader, or with a non-touch-sensitive rotary control, writing of automation begins and continues until you stop moving the fader.

Auto Latch mode works pretty much the same way, the difference being that the control doesn't return to its previously automated position when you let go of the control – it stays where you left it until you hit Stop. (However, at this point where you hit Stop, the automation value will change instantly to the previous value.) Latch mode is particularly useful for automating pan controls and plug-ins on non-touch sensitive rotary controls, since it does not time out and revert to its previous position when you release a control. If you complete your automation pass right to the end of the track, the control will stay wherever you last positioned it until the end of the track.

You can write automation on more than one track at a time by selecting an Automation write mode for each track that you want to automate.

tip ▷ If you are writing automation in Touch mode with Loop Playback enabled, automation will automatically stop being written when you reach the end of the looped selection. At the beginning of each subsequent loop pass you need to touch or move the control again to start writing new data.

Automation Playlists

Every Pro Tools track contains a separate automation playlist for each automatable control. You can choose which one of these to display in the Edit window by clicking on the Track View selector.

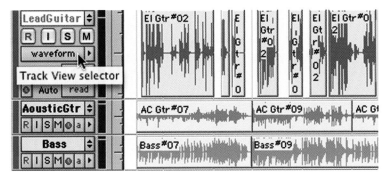

Fig. 10.23 – Clicking on the Track View selector.

This lets you choose between the Waveform or Blocks, Volume, Pan automation or Mute automation, Send automation parameters, or Plug-in automation parameters if you have any sends or plug-ins in use.

note ▷ Bear in mind that you won't see any plug-in automation parameters listed here unless you have also enabled these parameters for automation in the plug-in itself.

On Auxiliary Input and Instrument tracks the parameters include Volume, Mute, and Pan.

MIDI tracks work slightly differently and only offer automation for MIDI Volume, MIDI Pan, and Mute. Also, on MIDI and Instrument tracks you can display and edit other continuous MIDI controller data such as mod wheel, breath controller, foot controller, or sustain.

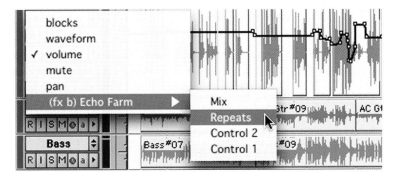

Fig. 10.24 – Choosing a display mode using the Track View selector popup.

There will be occasions during your mixing sessions when you will want to suspend the automation so that you can hear what is going on more clearly and make new decisions about the mix. You can suspend automation for the whole Session using the Auto Suspend button in the Automation Enable window, but, more often, you will want to suspend automation for specific controls. To do this, set the Track View selector in the Edit window to show the automation playlist for the control you want to suspend then Command-click (Mac) or Control-click (Windows) the control name in the Track View selector. To indicate that the control is suspended, its control name becomes italicized.

tip ▷ To suspend automation for all controls, Command-Shift-click (Mac) or Control-Shift-click (Windows) the name of any control in the Track View selector. To suspend automation for a specific control on all tracks, Command-Option-click (Mac) or Control-Alt-click (Windows) the name of the control in the Track View selector.

note ▷ Enabling and suspending any automation (other than Pan automation) from the Edit window affects all members of any applicable Edit Groups. You can override this group behaviour so that you only affect the track you are working with by Control-clicking (Mac) or Start-clicking (Windows) the control name.

Editing Automation Data

The Edit menu has four special commands (Cut Special, Copy Special, Paste Special, and Clear Special) that you can use to move or clear different types of automation data and MIDI controller data between tracks, sends, and plug-ins.

Using these commands you can just edit the automation and MIDI controller data without affecting any associated audio or MIDI notes. You can also edit automation and MIDI controller data without having to change Track Views, and the Paste Special command lets you copy one data type to another (e.g. Left Pan data into the Right Pan playlist).

The Cut Special and Copy Special commands let you cut or copy just the automation data from the current selection (without associated audio or MIDI notes) and place it in memory to paste elsewhere. Sub-menus let you choose between All Automation (which affects all automation or MIDI controller data whether shown or not), Pan Automation (which affects only pan automation or MIDI pan data whether shown or not), or Plug-In Automation (which affects only the plug-in automation that is currently being displayed).

When you have automation data in the computer's clipboard memory, the Paste Special commands become active. There are three choices in the sub-menu that let you paste the automation data into another region without affecting any associated audio or MIDI notes:

'Repeat to Fill Selection' pastes the automation data repeatedly until it fills the entire selection range, which is very useful for creating repeated drum patterns, for example.

tip ▷ Another situation where 'Repeat to Fill Selection' neatly solves a problem is when you have used the Region menu's Loop command to create repeated aliases of a region selection. The problem here is that the repeated regions do not include any automation that applied to the source region. To quickly sort this out, select the original automation in the source track and use the Edit menu's Copy Special command to copy the automation you are interested in (All, Pan, or Plug-in). Then use the Selector tool to highlight the loop aliases (or any other regions) that you want to apply this automation data to and use the 'Repeat to Fill Selection' command to paste the automation data to all the loop aliases.

'Merge' lets you add the pasted data to any existing automation data of the same type in the destination selection. You might use this, for example, if you like the pitch bend on a particular synth and want to use this on another synth.

'To Current Automation Type' pastes the automation data or MIDI controller data from the clipboard to the selection and, as necessary, changes the data to match the current automation or MIDI controller type. So, for example, you could copy Left Pan data into the Right Pan playlist. What you cannot do is to paste MIDI controller data to automation data or vice versa.

The Clear Special commands let you clear just the automation data from the current region – leaving any audio or MIDI note data untouched. Again, a sub-menu lets you choose between clearing All Automation or MIDI controller data (whether shown or not), only Pan Automation or MIDI pan (whether shown or not), or only the Plug-in Automation that is currently being displayed.

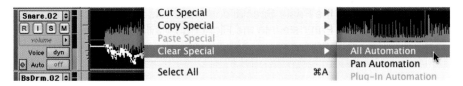

Fig. 10.25 – Using the Clear Special command to remove all the automation from the selected automation playlists.

Thinning Automation

When you create automation data in real time, Pro Tools creates many more breakpoints than you actually need. If the automation curve is relatively smooth, just a few breakpoints will suffice to represent your automation moves with a great deal of accuracy. Each breakpoint uses up some of the available random access memory (RAM) and CPU time and a point can be reached on a busy session where this impacts badly on the efficiency of your system.

To alleviate this situation, Pro Tools provides two different ways to thin automation data and remove unnecessary breakpoints: the 'Smooth and Thin Data After Pass' option and the Thin Automation command.

When the 'Smooth and Thin Data After Pass' option is selected in the Automation Preferences page, Pro Tools automatically thins the automation breakpoint data after each automation pass. You can also choose the amount of thinning that will be applied. The default setting of 'Some' works well in most situations, but you may still find it necessary to use the Thin command to remove even more breakpoints.

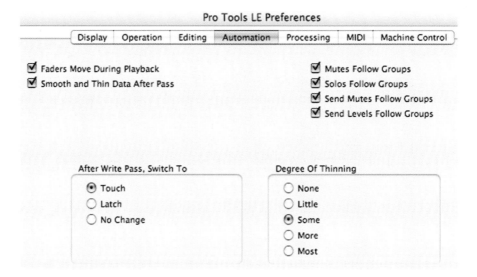

Fig. 10.26 – Automation Preferences with the 'Smooth and Thin Data After Pass' option selected and the 'Degree of Thinning' set to 'Some'.

The Edit menu's Thin Automation command lets you selectively thin areas in a track where the automation data is too dense. When you use the Thin Automation command, the degree of thinning that is currently selected in the Automation Preferences page is applied to the data, so you may need to change this. As with any trial and error process in Pro Tools, you can always use the Undo command if the thinning that you apply does not sound good when you audition the results.

To use the Thin Automation command:

Step 1. Click the Track View selector in the Edit window to display the automation type you want to thin.

Step 2. Choose the Selector tool and highlight just the automation data you want to thin, or use the Select All command to select all the data.

Step 3. Choose the Thin command from the Edit menu to thin the selected automation.

Creating Automation using the mouse

The simplest and most accurate way to create automation using the mouse is with the Grabber tool. When this tool is selected and you move the mouse over any displayed automation type in the Edit window, the cursor turns into a pointing finger. When you click on the automation line this inserts a breakpoint.

To remove a breakpoint you can Option-click (Mac) or Alt-click (Windows) on the breakpoint. If you want to remove several breakpoints it is quicker to change to the Selector tool and drag across the range of breakpoints you want to remove from the graph line to select these – then hit the Delete or Backspace key.

tip ▷

To move all the automation breakpoints in the track up or down while preserving their relative levels, choose the Trimmer tool then drag this up or down in the automation playlist at any point after the last automation breakpoint. To move the automation breakpoints in a particular section of a track up or down while preserving their relative levels, choose the Selector tool first and drag the mouse to highlight the range of interest. Then choose the Trimmer tool, point the mouse anywhere along the line of selected breakpoints and drag the whole line upwards or downwards.

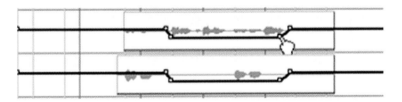

Fig. 10.27 – Inserting automation breakpoints using the Grabber tool.

You can also use the Pencil tool to create automation by drawing directly in the automation playlist. The default Freehand Pencil tool inserts a series of breakpoints as you drag the mouse in the automation playlist.

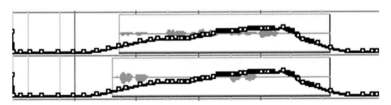

Fig. 10.28 – Drawing automation breakpoints using the Pencil tool.

If you click and hold the Pencil tool selector, a popup menu appears with options that let you draw straight Lines with a single breakpoint at each end; or Triangles that consist of a sawtooth pattern repeating at a rate based on the current Grid value with a single breakpoint at each extreme; or Squares with patterns that repeat according to the current Grid value; or Random patterns. The Triangle pattern can be useful to control continuous functions, and the Square pattern is typically used to control switched functions such as Mute or Bypass.

tip ▷ Since the Pencil tool draws these shapes using the current Grid value, you can use it to perform panning in tempo with a music track, or on frame scene changes when working in post-production.

The Final Mix

Having done all your preparation very thoroughly, the final balancing session can be relatively easy. In the early stages of your mixing session, you will have chosen most of the elements you want to work with and you will have set up the effects you want to use. It can take some time to set up sends to external effects units and Auxiliary tracks with chains of plug-ins to process the drums, the vocals, and the various instruments – especially if you are looking for that 'special' reverb sound or combination of delays, or if you are trying to find the right EQ settings and pan positions to make particular instruments 'sit' properly in your mix, or stand out as features. With a typical pop song, you should allow at least half a day to get everything more or less in place. Then you might take a break for half an hour or so before coming back to do the final balancing of levels and tweaking of effects. Here you should be concentrating on the most important tracks – typically the lead vocal and any featured solo instruments.

tip ▷ Don't forget how easy it is to loop playback of any section that you want to hear over and over again during your mix session: select a range of time by dragging the mouse through the timeline ruler and choose Loop Playback from the Options menu (or just hit Command-Shift-L (Mac) or Control-Shift-L (Windows) on your computer keyboard) then hit Play.

Make sure that all the tracks you are not using are hidden and made inactive, and set the rest to the minimum track height – leaving the lead vocal or solo instrument, and any other tracks that you want to tweak during your mixes, at medium or large track heights. It is also a good idea to display the Volume or other important Automation curves in the chosen tracks so that you can see clearly what is going on and so that you can manually adjust these at any stage during the mix session.

tip ▷ If you are familiarizing yourself with Pro Tools LE for the first time, it is worth taking a look at the Filtered Dream demo session that is supplied with the software to see how a session that has already been mixed looks. And don't forget that you can always clear the automation (an ideal opportunity to try out the Clear Special command) from the playlists in the Filtered Dream demo and make your own mix of this session.

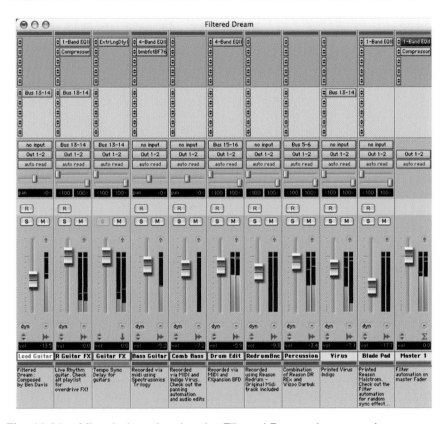

Fig. 10.29 – Mix window showing the Filtered Dream demo session.

If you are using the real-time automation features for your mix, you will start out in Auto Write mode. You probably won't get all your moves correct during the first pass, so you can use the Auto Touch or Latch modes to refine the sections that need changing.

Even if I start out using real-time automation, I usually end up going into the Edit window to manually edit the automation breakpoints to achieve exactly the right result. In this window it is so easy to see exactly where the vocal and instrumental phrases lie and to edit the automation curves to do exactly what you want them to do.

Take a look at the screenshot of the Filtered Dream demo session that is supplied with Pro Tools LE 7. I have arranged this with the MIDI tracks that were used to create the audio for the Trilogy Bass, Blade Pad and Redrum tracks hidden, and with the tracks set to sensible track heights with the Volume Automation playlists revealed.

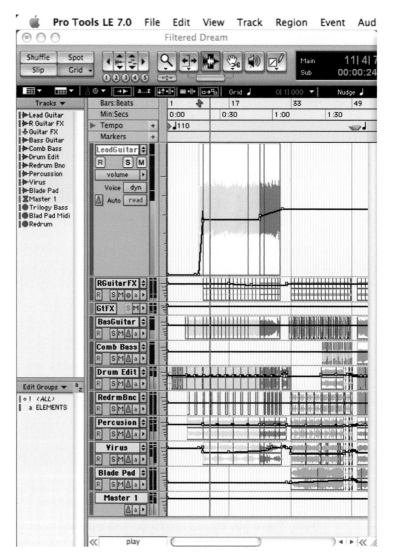

Fig. 10.30 – Edit window showing the Filtered Dream demo session ready to mix.

Mixdown

Once you have your mix sounding the way you want it, you will usually want to record this as a stereo 'master'. You have various options here. The first is to connect your main stereo outputs to a stereo mastering recorder such as a DAT machine, DVD- or CD-recorder, or even to a 1/2″ analogue tape recorder.

Alternatively you can create a stereo file on disk – either using the Bounce to Disk command in Pro Tools, or by recording the final mix to a pair of tracks in the same Pro Tools session – assuming you have these tracks available.

You can bounce to disk or record to new tracks anytime during your session if you want to 'print' (i.e. record) effects to disk so that you can free up your DSP to apply more plug-ins, or if you want to create submixes to use as loops or as 'stems' for post-production or whatever.

As the manual explains it, 'Printing effects to disk is the technique of permanently adding real-time effects, such as EQ or reverb, to an audio track by bussing and recording it to new tracks with the effects added. The original audio is preserved, so you can return to the source track at any time. This can be useful when you have a limited number of tracks or effects devices.'

Although you can hear the bounce as it is being created, you can't adjust any controls during a 'Bounce to Disk'. So you should only use 'Bounce to Disk' if you need to convert the bounced files or if you don't want or need to interact with any mixer controls during the bounce.

If you want to be able to adjust the mixer controls in real time while the files are being written, you should record to new tracks.

note ▶ Random access media (such as hard disks, optical disks or other computer data drives) can produce a true digital copy of your data, because every bit value is maintained. Sequential media (such as DAT tapes) use error correction schemes to fix the occasional bad data that is received in a digital transfer. These corrections are deviations from the actual data, and with successive reproductions, represent a subtle form of generation loss. You can avoid this loss by creating and maintaining your master mixes on random-access digital media (such as a hard drive) and transferring them to sequential digital media (such as DAT tapes) only as needed.

Recording to a Stereo Mastering Recorder

Using the main stereo analogue line outputs from your Pro Tools interface, you can record your master mixes onto analogue mastering recorders, such as 1/4″ or 1/2″ tape recorders. In this case there is no need to convert sample rates or bit-depths or to add dither, as the D/A converters will convert the digital audio to analogue audio (with very little loss of quality) no matter which sample rate or bit-depth you are using.

Although it is usually best to create digital master mixes as files on hard disk, Pro Tools also lets you record your master mixes digitally, direct to any AES/EBU-equipped or S/PDIF-equipped digital recorder such as a DAT or CD-Recorder. Note that these are 16-bit digital systems, so if you have been working at 24-bit resolution, you will inevitably lose some quality.

note ▷ Pro Tools LE systems support 24-bit input and output and use 32-bit float-ing point mixing and processing internally. This allows you to exchange digital audio data with other 24-bit recording systems without any bit-depth conversion.

Bounce to Disk

The Bounce to Disk command writes the current session (if you have not made any selection), or the Edit or Timeline selection, as new audio files to disk. Any available output or bus path can be selected as the bounce source and sample rate, bit depth, and other conversion processes can be applied during or after the bounce.

Basically, you mute everything but the tracks you want to bounce, make sure that all the levels, pans and any effects and automation are the way you want them to be, assign the outputs from all the tracks to the same pair of outputs, then select the Bounce command from the File menu to open the Bounce dialog.

note ▷ Audio coming from external inputs cannot be recorded during a 'Bounce to Disk', so if you need to include any external inputs in your bounce, these must be recorded to new audio tracks first. Also, when Low Latency Monitoring is enabled, only audio tracks are included with the Bounce to Disk command – Auxiliary Input and Instrument tracks are ignored. To include these, you need to disable Low Latency Monitoring before the Bounce to Disk.

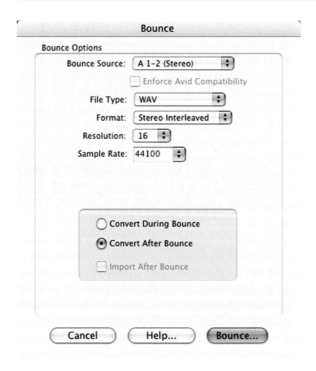

Fig. 10.31 – Bounce to Disk dialog.

In the Bounce dialog, the first popup lets you choose the Bounce Source. This source can be taken from any of the audio outputs or busses actually in use in your system.

note ▷ Don't forget to choose your bounce source to match the output assignments on your session. The source selection defaults to outputs 1/2 and you may be using a different output pair in your session. If you do forget, there will be no audio in your bounced file.

The File Type popup lets you select from Broadcast WAVE (BWF), Audio Interchange File Format (AIFF), Sound Designer II (SDII), MPEG-1 Layer 3 (MP3), QuickTime, Sound Resource (or MXF format if you have DigiTranslator installed).

note ▷ MXF (Material Exchange Format) files include both video and audio data, and are designed for interchange of audio-visual material with associated data and metadata. Advanced Authoring Format (AAF) and Open Media Framework (OMF) sequence format files can refer to MXF media files, or have MXF media files embedded within them. Pro Tools with DigiTranslator supports AAF embedded sequences.

The MPEG-1 Layer 3 compression format (MP3) is used for streaming and downloading audio over the Internet, and for playback on portable devices. The MP3 encoder provided as an install option with Pro Tools works for 30 days after which you are expected to buy it from the online DigiStore at www.digidesign.com.

Pro Tools systems running on Windows can also create Windows Media format (.wma, .wmv, or .asf) files containing audio, video, or script data.

File format options include summed mono, multiple mono, and interleaved stereo. If you intend to import the files back into Pro Tools you should choose mono or multiple mono files. Interleaved stereo files can be used by other software such as BIAS Peak which you may use for any final edits or processing, or by CD-burning software such as Toast or Jam. Resolutions available include 8-bit for multimedia work, 16-bit for CD distribution and 24-bit for high-quality digital audio systems. All the standard sample rates from 44.1 up to 192 kHz are available.

You can also choose whether to convert during or after the bounce. I often choose to convert after the bounce even though this takes longer, as it leaves the processor free to concentrate on one task at a time – bouncing then converting. You can select the option to Import After Bounce if you want to use the new tracks in your session after the bounce – or leave this unchecked if you simply want to create master mixes on disk that you will assemble and check later.

When you hit Bounce, you will be presented with a Save dialog box where you can name your new file(s) and choose where to save these.

The Bounce to Disk command uses all the available voices from your session and all audible tracks will be included in the bounce – including tracks that 'pop through' when other tracks are not using their voices. Muted tracks do not appear in the bounce. All the read-enabled automation is applied along with all plug-ins that are in use and any processing that is being applied via hardware inserts.

Bear in mind that the bounced mix will be exactly the length of any selection you have made in the Timeline or Edit window – which should be linked. If you want to include any reverb trails or other effects which 'hang over' at the end of the track you will need to select additional time to accommodate this. If you don't make any selection, the bounce will be the length of the longest audible track in your session.

So when should you use Bounce to Disk and when should you record your mix to new tracks in your session? If you have totally automated your mix, or if you have no free tracks available in your session, then Bounce to Disk is the best option. However, some engineers may still prefer to do a manual fade at the end or to tweak faders or other controls on-the-fly during the final mix-down. In this case, recording to new tracks or to an external recorder are the best ways to go.

Recording to Tracks

Recording your mix to new audio tracks is just the same as recording any other input signals into Pro Tools. Obviously, you need to have sufficient free tracks, free voices and bus paths available. The beauty of this technique (compared with Bounce to Disk) is that you can add live input to your mix or adjust volume, pan, mute, and other controls during the recording process.

Once you have your mix set up with the levels, pans, plug-in processing, and routing all sorted out, you are ready to record your new tracks containing your mix, submix, or stem. Simply record-enable the new tracks, click Record in the Transport window, then click Play in the Transport window to begin recording – which will begin from the location of the playback cursor. Recording will continue until you press Stop or punch out of recording – unless you have selected a particular section.

It can help save a little time later if you select exactly the length of audio that you want to record to these new files. Even if you place the playback cursor exactly at the start of the audio you are unlikely to be able to stop recording at exactly the right moment. To make a selection, link the Edit and Timeline selection using the command in the Operations menu and drag the selection cursor over the length you want to encompass.

Don't forget to select some extra time at the end of your selection to accommodate reverb tails, delays, or any other effects that may still be sounding after the audio has finished as the new recording will stop automatically immediately at the end of your selection.

And don't be too keen to select too close to the start point either. It is always better to include a bit more than you think you will need in your selection – you may need to edit some 'room tone' that you find there to fill a gap somewhere else, or to analyze some noise that is there so that you are better able to extract this noise from the rest of the session.

When to Use Dither

Whenever you change the bit-depth of digital recordings you need to apply dither to reduce quantization error that can become audible, particularly when fading low-level signals. Dither does this 'trick' by actually introducing very low-level random noise that, counter-intuitively, increases the apparent signal-to-noise ratio. Typically, you will apply dither if you are bouncing a mix to a file on disk that uses a lower bit-depth than your session.

Two plug-ins are supplied for this purpose – POW-r Dither and Dither. The POW-r Dither plug-in offers 16-bit and 20-bit resolutions along with three choices for Noise Shaping – Types 1, 2, and 3. The Dither plug-in has 16-, 18-, and 20-bit options to cater for all possible scenarios but only features one standard Noise Shaping type.

note ▷ Noise Shaping shifts the noise introduced by the dithering process to frequencies around 4 kHz where this low-level noise is likely to be masked by the audio program material.

If you are mastering from a 24-bit session to a 24-bit digital recorder or to analogue tape via 24-bit D/A converters, there is no need to apply dither. The 20-bit option is provided for compatibility with some digital devices that use this format. On the other hand, if you are mastering to a 16-bit medium – whether this be a file on disk or an external recorder – you should apply dither. You may be seduced into thinking that if the original session is 16-bit you don't need to dither to another 16-bit medium – but you would be wrong. Although 16-bit sessions save their data to 16-bit files, they are actually processed internally while the session is running at higher bit rates – 32-bit floating point for Pro Tools LE systems. So it doesn't matter whether you are running a 16-bit or a 24-bit session – you should still dither when mastering to 16-bits.

tip ▷ Dither is not automatically applied when you use the Bounce to Disk command – so you need to insert and apply the Dither plug-in on your Master Fader before your bounce if you want to create a dithered file. Bear in mind that if you do not apply dither and you choose to convert to a lower resolution, say from 24-bit to 16-bit, during or after a Bounce to Disk, the resultant file will be converted by truncation that is the low-order bits will simply be 'thrown away' and quantization noise may become audible.

What you have learned

After completing this chapter you should know much more about how to set up your monitor speakers, what monitoring levels to use and how to meter levels in Pro Tools.

You will be aware of how to set up Master Faders and use multiple output assignments.

You will also be aware of how you can make tracks inactive and why you would want to do this, how to group tracks, and how to use the track Output windows.

There are many ways that you can use Auxiliary tracks and Sends to route audio around the Pro Tools mixer. You should now be aware of the different options available for yo to use.

Automation is another important facility that is available while mixing. You should now be familiar with the features that Pro Tools offers.

This chapter also explains the pros and cons of bouncing tracks versus recording tracks to disk when making your 'master' mixes.

Backups

When you have finished your project you should tidy this up and back it up to make way for the next project.

A golden rule of computing says that if it only exists on one hard disk, it probably doesn't exist at all. All it takes is a few errors on the hard drive to make the disk directory unreadable, so that you can't find your files. Software such as Disk Warrior can help prevent such catastrophes, but disasters can still happen. As I am writing this chapter, I am having disk directory problems with three of my hard drives, and one appears to have completely failed. And these drives are all less than 18 months old and have been regularly maintained. One of the drives was purchased just a week ago!

Of course, if a disk drive completely fails, you stand a good chance of losing all your files permanently. There are commercial data recovery services available, but these usually charge 'an arm and a leg' and can never guarantee to recover all, or even any, of your data.

I have two identical external drives so that I can keep a copy of the data from the first drive on the second drive at all times in case one of these drives fails. When I have completed a project, and sometimes at various stages throughout a project, I make backups onto CD-R or DVD-R.

Before backing up, you should thoroughly tidy up your session. If you need to work on the session at any time in the future, you will save a lot of time if you have left it in a tidy state. Also, if you intend to take a project to another studio to do more recording, editing, or mixing on another system, a tidy session always makes transfers run more smoothly. For example, you should remove any unused regions and files, and make sure that all your regions and files are named sensibly.

When this is done, use the Save Copy In. . . command to save a copy of the Session file along with copies of all the audio files used in the project. You can then back this up to CD-R or DVD-R secure in the knowledge that all the necessary audio files are included in the backup along with the Session file.

You can also transfer the copied session to any other system, again confident in the knowledge that all the necessary audio files are with the Session file.

Transfers

Preparing Pro Tools Sessions for Transfer

Typically, after a busy Pro Tools session you may have several tracks containing alternate takes that you subsequently decide will never be used. You might also have other tracks that you used for one reason or another along the way that are no longer needed. Before making your transfer, you should save an edited version of your session with these extra tracks removed.

It is also quite likely that you have a number of unused regions clogging up your Region List as a result of trial edits, false takes, and so forth. The quickest way to get rid of most of these is to choose 'Select Unused Regions Except Whole Files' from the popup menu at the top of the Regions List.

You should choose the 'Except Whole Files' option because otherwise any whole files that you had edited to create shorter regions from would become selected – which, of course, you would not normally want (It is useful to retain the region that encompasses the whole file in case any adjustments need to be made when editing further 'down the line').

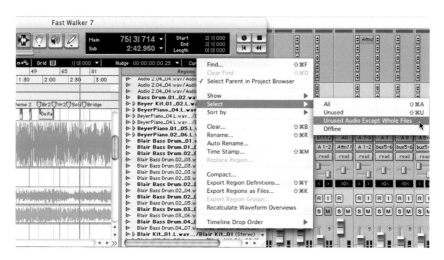

Fig. A1.1 – Selecting unused files.

The next step is to choose the 'Clear' command from this menu, which brings up the 'Clear Regions' dialog. Here you can choose whether to simply remove the selected regions from the list or whether to completely delete these from your hard drive. You should always choose the 'Remove' option unless you know for sure that all these unused regions belong to files for which you have no further use.

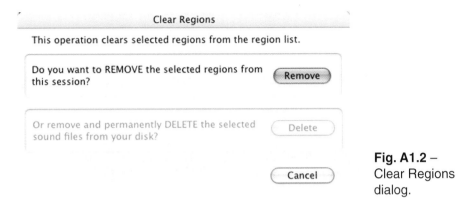

Fig. A1.2 – Clear Regions dialog.

You should also carefully go through the whole file regions, identified in bold type in the list, to get rid of any false takes or other stuff that you will never need again – auditioning each of these carefully first if you are at all unsure. When you identify any such files, you can use the 'Clear Regions' command to get rid of these, choosing the 'Delete' option this time to remove them from your hard disk.

When you have tidied up your session, your next move should be to make a copy of the session – complete with any associated files. Choose the 'Save Copy In. . .' command from the File menu.

In the 'Save' dialog that appears, you can choose the sample rate, bit depth session format and audio file type for the target system to which you are making your transfer.

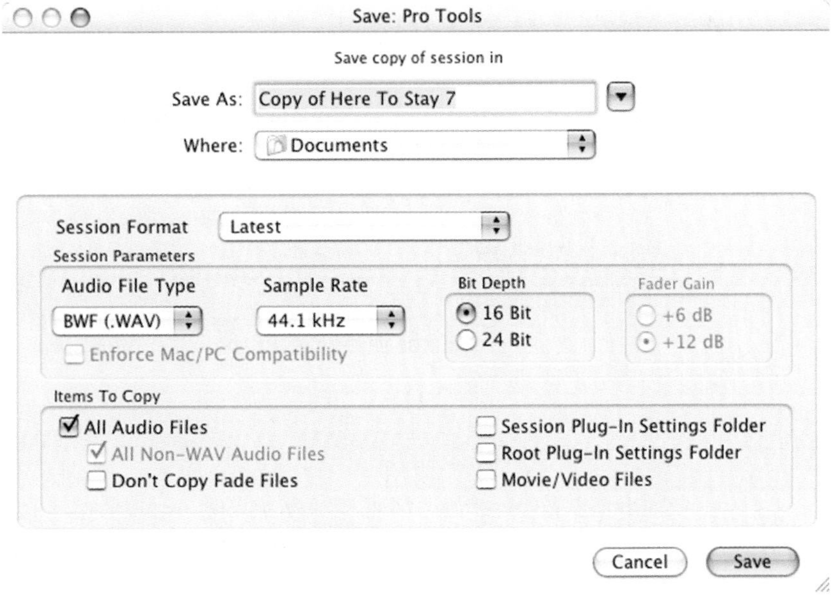

Fig. A1.3 – Save Copy In. . . dialog.

You should select the box for 'All Audio Files' to make sure that all the audio files associated with your project will be copied – even any files that may exist on a different hard drive or in a different folder on the same hard drive. You can also include various other files and settings as necessary.

Types of Transfers

You may be moving from a Pro Tools LE system to another LE system with different plug-ins and a different audio interface, or to an M-Powered system or a Pro Tools|HD system. You may have started your project on a PC running Pro Tools LE at home and you may wish to transfer to a Pro Tools|HD system running on a Mac at the professional studio. It is also possible that you want to transfer from Pro Tools to MIDI + Audio sequencer (such as Cubase SX, Nuendo, Logic Pro, Digital Performer, or Sonar) or vice versa. And if you are working to picture, you may want to transfer a project into Pro Tools from Avid Xpress or from Final Cut Pro and back again.

Moving Pro Tools Sessions between Systems

The good news is that Pro Tools is extremely flexible when it comes to moving sessions around between different systems. The software will automatically deactivate tracks, routings and plug-ins not available when moving to a different system – while retaining the data for use when transferring back. Unique file identifiers help to resolve file and location references when moving between systems. Pro Tools also supports work with offline media, allowing a session to be opened and edited even if all the session's audio or video files are currently unavailable – and any edits that you make to tracks containing offline media are reflected in the session when the files are available again.

Also, to conserve DSP resources in a session, tracks, I/O assignments and plug-ins can be set to 'inactive'. Inactive items retain their various settings, routings, and assignments, but are taken out of operation – freeing the DSP they were using for other uses. The original settings will remain saved so you can always see what you've deactivated and return to these at any time. Even better – when you move a Pro Tools session to a system that has different plug-ins and I/O configurations, Pro Tools will automatically deactivate tracks, plug-ins, sends, or I/O channels as necessary while letting you preserve your original session settings so you can return to these when you move back to the original system.

Mac/PC Compatibility

All newly created Pro Tools 7.0 Sessions are compatible on both Macintosh and Windows systems. Older versions of Pro Tools are not necessarily Mac/PC compatible, but an 'Enforce Mac/PC Compatibility' option is available in the 'Save Copy In' dialog to let you save sessions to be compatible with lower versions of Pro Tools. This ensures that all Pro Tools files are saved in the Audio Interchange File Format (AIFF) or Broadcast Wave (BWF) (.WAV) format, and that no illegal characters are used in file names.

note ▷ If you create a Pro Tools 6.9.x or lower session on a Macintosh system, the session will only be compatible on Windows systems if the 'Enforce Mac/PC Compatibility' option is selected when using the 'Save Copy In...' dialog.

The file format to use for cross-platform interoperability is BWF. This uses the. WAV file extension and is now the default format for new Pro Tools sessions. Older Mac versions used the Sound Designer II (SDII) format but this is not supported on Windows and only supports sample rates up to 48 kHz. AIFF format can also be used on both platforms, but lacks the metadata features of the BWF.

You should also use 3-letter file extensions for cross-platform compatibility. The Mac does not insist on this, but Windows requires that all files in a session must have a 3-letter file extension added to the file name. Make sure that any Wave files created on the Mac have the '.wav' file extension, that any AIFF files have the '.aif' file extension, that Pro Tools 5.1 to 6.9.x session files have the extension '.pts,' and that Pro Tools 5 sessions have the extension '.pt5.'

note ▷ There are restrictions on the ASCII characters that you use to name regions, track names, file names, and plug-in settings if you want to achieve full Mac/ PC compatibility. As you might expect, any character typed with the Mac's Command key won't be available for Windows. Less obviously, you should avoid the following:

/ (Forward Slash)
\ (Backslash)
: (Colon)
* (Asterisk)
? (Question mark)
" (quotation marks)
< (Less than symbol)
> (Greater than symbol)
| (Vertical line or pipe)

When you import files into a session, incompatible characters are converted to underscores ('_') and the renamed files are placed in a folder called 'Renamed Files'.

Transferring Files between Different Computer Platforms

There are many ways to get your Pro Tools files from PC to Mac or vice versa. If the two computers are networked, you can transfer the files using Ethernet.

Then there is always 'sneakernet', where you put the files onto a CD-ROM or DVD-ROM and take this to the other computer. Just make sure that the CD-ROM is prepared with the correct options to work on both Mac and PC. For example, Toast 7 for the Mac offers 'Mac Only', 'Mac & PC', 'DVD-ROM', and 'ISO 9660' formats. Obviously, you need to avoid the 'Mac Only' option if you want to take your files to a PC.

CD-R gives you the best chance of compatibility with the widest range of target systems. DVD-R is a good choice if you know that they are going to a high-end Mac or PC that has a built-in DVD drive. Any other removable disk or tape or hard drive system would work just as well – but not everyone has Sony AIT or Quantum Digital Linear Tape (DLT) drives, for instance.

You can copy your Pro Tools files directly from an New Technology File System (NTFS) (Windows XP) drive to an HFS+ (Mac) drive (or vice versa) on a Windows XP system with the MacDrive utility installed.

tip ▷ MacDrive lets you plug a MacDrive into a PC FireWire or USB port without having to reformat the drive. You can get hold of MacDrive 6 for Windows from MediaFour at www.mediafour.com/products/macdrive6/.

For example, a Windows XP session stored on an NTFS formatted drive can be copied to an HFS+ formatted drive connected to the Windows computer. The MacDrive 6 Utility has options for Backup/File Transfer. With this set, you can simply drag the session folder from the NTFS drive to the HFS+ drive.

If you want to go the other way, to transfer Pro Tools Macintosh sessions from HFS+ drives to NTFS drives, set the MacDrive 6 Options to Normal Use then drag the session folder from the HFS+ drive to the NTFS drive.

note ▷ If you transfer a Pro Tools session from the Mac that uses Windows-incompatible characters, the Pro Tools sessions will be unable to re-link to the unfortunately-named audio files and fades in Windows. The illegal characters will be converted to underscore ('_') characters and these files will be saved to a folder that is automatically created, called 'Renamed Audio Files'. You will then have to manually re-link each file by File ID, which can be a tedious process. The best way to avoid this is to avoid using illegal Windows characters in your Mac sessions in the first place.

Moving between Pro Tools | HD and Pro Tools LE or M-Powered systems

One of the most useful features in Pro Tools is the session interchange facility provided between Pro Tools | HD systems and Pro Tools LE or M-Powered systems.

Pro Tools | HD 7 sessions can have up to 256 tracks while LE and M-Powered 7 only support 128 tracks, and only 32 of these can be played at the same time. Also, Pro Tools | HD 7 systems support up to 128 Instrument tracks, 128 busses and 10 sends while Pro Tools LE or M-Powered 7 systems only support up to 32 Instrument tracks and mix busses – and older Pro Tools versions only support up to 16 busses and 5 sends.

So what happens if you open a Pro Tools | HD 7 session with more than 32 mono tracks, Instrument tracks or mix busses into Pro Tools LE or M-Powered 7? Specifically, for any tracks beyond the first 32, as well any inactive tracks, the track's Voice Selector is set to 'voice off'. Also, any Instrument tracks and any bus assignments beyond 32 are made inactive and any sends beyond the first five are removed.

Pro Tools LE and M-Powered systems do not support multichannel surround tracks, so any of these will be removed. Similarly, Pro Tools | HD 7 systems may have many more inputs and outputs than LE or M-Powered systems. If this is the case, any input and output paths that are not available are made inactive.

If you are using plug-ins when exchanging project files between Pro Tools platforms, ideally you will want to have the same plug-ins installed on all systems involved. You can get around this by bouncing tracks that have plug-in processing to audio tracks that include these effects 'printed' to disk. And if your studio tracks use DSP-based TDM plug-ins, you can bounce these to disk before transferring to your Pro Tools LE or M-Powered system.

Nevertheless, the way that Pro Tools LE and M-Powered handle sessions that use TDM plug-ins is particularly neat – TDM plug-ins with Real-Time AudioSuite (RTAS) equivalents are converted; those without equivalents are made inactive. As you will appreciate, you can't use TDM plug-ins with LE systems – so you could end up losing lots of useful signal processors that you have used to build your session. Fortunately, there are now RTAS equivalents for many TDM plug-ins, so if you have used any of these in your TDM session, all the plug-in settings will transfer to the LE system – where it will use the RTAS equivalent. If there is no RTAS equivalent, the settings will be transferred to the LE version, but the plug-in will be made inactive. When you transfer this back to the TDM system the TDM plug-ins will become active again and the settings data will still all be there.

note ▷ This even works when transferring between two differently configured TDM systems – Pro Tools will automatically deactivate any unavailable routing assignments and plug-ins while keeping all settings and automation data (including references to plug-ins not installed on the current system).

tip ▷ If you have used more tracks, busses or sends in your Pro Tools│HD 7 session than your LE or M-Powered system can handle, then it is a good idea to
make some decisions before making your transfer about how to handle
these. One possibility is to save a version of your Pro Tools│HD 7 session
that only includes the 32 most important tracks. Another way is to bounce
combinations of tracks together, such as a stereo mix of all the drums, to get
the track count down to 32 before you make the transfer. You could even
bounce everything as a stereo mix and add a further 30 new tracks on your
LE system – then transfer these back to the Pro Tools│HD 7 system and combine the two sessions using the Import Track feature. The Import Track feature will seamlessly transfer audio and MIDI tracks between sessions with all
the mixer settings, tempo maps, plug-ins, and automation intact, so you could
open either the original Pro Tools│HD 7 session and import the new tracks
from the transferred LE session – or vice versa.

Transferring Pro Tools 7 Sessions to earlier versions

You may also need to transfer sessions from Pro Tools 7 format to Pro Tools
6.9 or lower. You can choose to save sessions in earlier formats using the
'Save Copy In. . .' dialog, which can be accessed from the File menu.

When you save session to versions 5.1 to 6.9 formats, Fader Gain levels and
automation breakpoints higher than +6 dB will be changed to +6 dB, any
Instrument tracks will be split into separate Auxiliary Input and MIDI tracks,
and any long names will be shortened to 31 characters. Also, any Region
groups or loops, any marker locations above 201, any buses above 17, and
any sends F–J with any associated automation will be dropped and the attributes marking any MIDI regions or tracks as sample based will be dropped.

If you save sessions in version 5.0 format, any multi-mono plug-in instances,
multichannel tracks, or sends assigned to multichannel paths or sub-paths of
multichannel paths will also be dropped. In addition, any tracks assigned to
'No Output' will be routed to Busses 31 and 32; any tracks or sends assigned
to Busses 33–64 will be routed to Busses 31 and 32; any tracks assigned to
multichannel paths or sub-paths of multichannel paths will be routed to
Busses 31 and 32; and any tracks or sends assigned to stereo paths referring
to even/odd channels (such as 2–3) will be routed to Busses 31 and 32.

Working with offline media

Now what if you don't have enough hard disk space to hold the audio and
video files that you are working on in the studio – but you want to do some work
at home using MIDI, for example? No problem here, as Pro Tools will allow
you to open and edit sessions even if none of the audio or video files used in
the original sessions are available.

Regions are said to be 'offline' when their parent audio files cannot be located,
or are not available, when you open a session or import a track. Offline regions
appear in playlists in the Edit window as light blue regions with italicized names.

Also, in the Region List, offline region names are italicized and the listed names are dimmed.

Here's the trick: although the region is offline, you can still edit the offline region – making cuts, deleting sections, copy or repeating sections – and when the media is available again, that is, 'online', these edits will apply to the online media just as though the media files were there when you made the edits. Of course, this only goes so far – you can't apply AudioSuite processing to an offline region for instance – but it does mean that you can work on your session much more flexibly using this feature.

Transferring Projects between applications

There are two ways to transfer projects between MIDI + Audio applications – the hard way and the easy way! Which way you will be able to use will depend on which software versions you are using. With older software versions of these applications, earlier than, say, the summer of 2003, it is much more likely that the only way is the hard way. Thankfully, with most of the software released since July 2003, you can probably do it the easy way.

The Hard Way: Using a MIDI file and separate audio files

Transferring projects from one MIDI + Audio workstation to another can be a tedious business. Nevertheless, with a few tips and hints to help you through, it can go quite smoothly.

The MIDI File format makes a reasonable job of transferring MIDI sequences from one MIDI sequencer to another, complete with tempo information, markers, and track names. And MIDI files can easily be transferred across platforms – from PC to Mac, for example. Nevertheless, it can still take some time to re-assign the track outputs on a different system or platform. And unless the target system has access to the same synthesizers and samplers loaded with the same selection of patches and samples it can be very difficult to recreate the original sounds. Synthesizer patches can be saved via SysEx and transferred as computer files, and samples can be transferred on disk or across a network to a new system – but this all takes time and trouble.

tip ▷ A better way is to record any MIDI tracks as audio before transferring the project. This way, what you were hearing before the transfer should be what you hear after the transfer.

note ▷ You should make sure that any audio tracks that you want to transfer start at the beginning of your sequence (Bar 1, Beat 1, or wherever). If you have any audio in your project that does not start at the beginning of the sequence, then you need to use your software's 'bounce to disk' feature to create new audio tracks that do start at the beginning of the sequence and that end just after the last audio finishes on each track – taking into account any reverberation or effects that may need time to die away at the end of the tracks.

Quite often, people build up musical arrangements using audio files containing short sections of music (audio regions) that only play at certain positions within the arrangement. You could write down the Bar, Beat, Clock positions and the lengths for each audio region within, say, a Cubase SX project and then painstakingly recreate these in, say, Pro Tools – but this could take a very long time for a large, complex project. That's why the only practical solution is to bounce each track (that consists of multiple regions or that doesn't start at the beginning of the sequence) to disk, starting at the beginning of the sequence and finishing after the last piece of audio has finished sounding. This is the tedious bit. However, once this is done, it then becomes trivially easy to transfer the project: just open a new file in Pro Tools (or whichever), import all the audio files from Cubase SX (or whichever), and place these in new audio tracks starting at the beginning of the sequence.

The Easy Way: Using a MIDI and OMF, AES31, or AAF files

Support for OMFI, BWF/AES31 files, and the new AAF format, means that you can now swap projects between Apple Logic Pro, Yamaha's Steinberg Nuendo and Cubase SX, Digidesign Pro Tools and Avid video systems – whether running on Mac or PC – with relative ease.

These formats contain information about where all the audio regions are positioned within your project – so you don't need to go to the trouble of making all your audio regions start at the beginning of the sequence before making transfers.

Open Media Framework Interchange (OMFI), often referred to simply as OMF, was originally developed by Avid and others to allow interchange of video and audio files between workstations, complete with their edits. The OMF format is supported by Pro Tools (with the DigiTranslator option), Nuendo, Cubase SX, Logic Pro, Digital Performer, and Sonar.

AES31 was developed as an interchange format by the Audio Engineering Society that will take information about the positions of events, fades and so forth. AES31 uses BWF as the default audio format.

The AAF is a relatively new interchange format, available since 2001, sponsored by the AAF Association (a large group of professional media manufacturers) – see www.aafassociation.org. AAF works much like the OMF interchange format – allowing content creators to easily exchange digital media and metadata across platforms and between AAF-compatible systems and applications.

The metadata incorporated into OMF and AAF files includes info about each media file, such as sample rate, bit depth, region names, the name of the videotape from which the media file was captured, and even time code values that specify where a file was used in the project. It also includes information about what files are used, where they appear in a timeline, and automation. For AAF or OMF sequences, information about unrendered AudioSuite effects (such as real-time EQ) on Avid workstations is also included. (Rendered effects are media files that can be imported or skipped on import into Pro Tools, and Pro Tools always skips unrendered effects on import.) And for AAF or OMF sequences, information is included about automation (clip-based gain or keyframe gain).

If your software supports any of these formats, this greatly simplifies the process of transferring the audio between projects, although you still have to use a MIDI file to transfer any MIDI data.

note ▷ There are two ways to handle media files when exporting OMFI files: using embedded media or using external file references.

Exporting to OMF with embedded media results in one large file containing both the metadata and all associated media files – with an upper file size limit of 2 GB. However, DigiTranslator only supports AAF export with external file references – it does not support AAF files with embedded media.

When you export projects using AAF or OMF with external file references this produces a metadata file with the '.aaf' or '.omf' file extension that simply contains references to the associated media – which remain untouched, stored as separate files. One advantage here is speed of operation – you are not, necessarily, having to copy the media files. And you are not likely to encounter any problems with file sizes on large projects, either. The main disadvantage is that you may have a large number of associated media files and all these have to be transported between systems. In this case, the potential for overlooking a file or whatever is obviously much greater.

Preparing Projects for Transfer

Before exporting an OMF or AAF file from your project, you should tidy this up, back it up, then prepare a special version for transfer to the target system.

If you are preparing a project for transfer to someone else's studio, then you should ask whoever will be engineering the project to specify the format for the audio files you will supply, such as Wave or AIFF. And it would be wise to make sure that the transfer works properly by checking it out at their studio at least a day before it is actually needed – to allow time for any errors to be corrected.

tip ▷ It makes sense to record your project using the sample rate and bit depth that matches that of the target system – otherwise you will have to convert your files at some stage or other, which takes time and can lose quality.

note ▷ If you export an OMF file from Pro Tools using sample rate conversion and import it into a project with a higher sample rate (for example, from a Pro Tools session at 44.1 kHz to a media composer project at 48 kHz), clicks will appear in the audio at region separation points and at the beginning of fade outs and end of fade ins (but not in crossfades). To avoid this problem, export the OMF file from a Pro Tools session with the same sample rate as the destination project.

Checklist

1. Delete any tracks and any audio or MIDI regions and files that you don't intend to use again.
2. If the project uses any MIDI or Instrument tracks, you can either record these as audio or export the MIDI data as a MIDI File.
3. If you have tracks or regions that use automation, it often makes sense to bounce these to disk first.
4. Similarly, if you want to export audio with the sound of any plug-in effects applied, you should process your audio using these first. Either create new files using AudioSuite plug-ins or bounce or record to disk if you are using RTAS or TDM plug-ins. Bear in mind that the target system may not have the same plug-ins available as the original system.
5. If your project uses more tracks than the target system can handle you may need to reduce the number of tracks – making submixes if necessary.
6. Use the Save Copy In. . . command, choosing the options to copy all audio files and any other needed files and settings, to save the specially prepared version of your project that you will use to transfer. You can also choose the required file formats, sample rate and bit depth for the target system here.
7. Finally, you should create a text file containing info such as the tempo and any special instructions for your project that you can keep with the transfer files.

OMF and AAF

If you have installed DigiTranslator, Pro Tools lets you export (or import) individual tracks or an entire Pro Tools session in OMF format or AAF format. These formats preserve all the edits you have made within your session and make transferring projects to and from other MIDI + Audio software applications much easier than before these formats were developed.

note ▶

> In earlier versions of Pro Tools, the DigiTranslator option was contained within the Pro Tools software as a demo version that could be authorized using an iLok key.
>
> Digidesign 7.0 plug-ins and Pro Tools 7.0 software options no longer include time-limited demos. Instead, plug-in and option demos require an iLok USB Key and an iLok license for evaluation. To obtain a demo license, go to the individual product pages located on the Digidesign Website (www.digidesign.com) and click on the Demo button.
>
> When DigiTranslator is installed and authorized, the File Menu command, 'Export to OMF/AAF' becomes active to let you export projects, and the 'Open Session' command allows you to import projects in OMF format.

OMFI format (usually referred to as OMF format) has been around for a long time in the video world and has become widely available for MIDI + Audio software – so this makes a good choice for project transfers. The more recently developed AAF also allows projects to be transferred between Pro Tools and AAF-compatible systems, but is not as widely implemented in other MIDI + Audio software. Both of these file formats can either contain links to the associated audio files or can contain audio files actually embedded (contained) within the OMF or AAF files.

DigiTranslator is available for both Pro Tools | HD and Pro Tools LE systems. With this installed, you can import AAF or OMF sequences into Pro Tools using either the Import Session Data or Open Session commands. You can also drag and drop AAF or OMF sequences or audio files from any DigiBase browser.

note ▷ Although Pro Tools can open and import AAF sequences that refer to other media files, AAF sequences that contain embedded media cannot be imported into Pro Tools. Also, if you plan to perform perf-based edits with a film project in Avid Xpress Pro, you should capture all media as OMF, and choose OMF instead of AAF when exporting a sequence for use in Pro Tools. Pro Tools does not recognize perf-based edits in an AAF sequence, and will place the audio at the full frame boundary.

Exporting AAF and OMF from Pro Tools

You can export individual tracks or an entire Pro Tools session in OMF format or AAF format using the Bounce to Disk, Export Selected as Files, and Export Selected Tracks as OMF commands. You can also export AAF and OMFI sequences using the Export Selected Tracks as OMF/AAF command.

note ▷ AAF sequences and OMF sequences and files exported from Pro Tools only retain time code addresses, region names, and definitions. They do not support video files, or retain information about plug-ins' assignments or parameters, routing, or grouping. Digidesign recommends that you bounce or record to disk any tracks that use effects prior to export.

When you select the tracks you want to export and choose 'Export to OMF/AAF' from the File menu, dialog appears that lets you choose the file formats and set various other options.

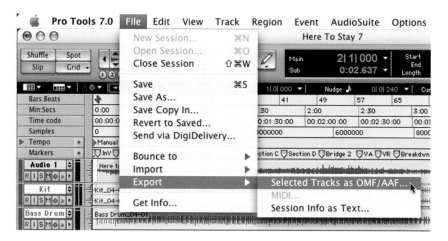

Fig. A1.4 – Exporting selected tracks.

The Export to OMF/AAF dialog defaults to exporting AAF format files, even though it mentions OMF files first in its name. So make sure to change the export format to OMF if this is what you want to use.

Also, if you want to export the session's volume and pan automation to AAF and OMF, you must make sure that the 'Quantize edits to frames boundaries' option is disabled (not selected). This option is provided for compatibility with Avid systems.

Fig. A1.5 – Selecting OMF format.

To avoid using more space on your hard drive than is absolutely necessary, the OMF or AAF files that you create by default can be linked to the source media – the original audio files – instead of creating new copies of the source media. This is fine if you want to transfer your Pro Tools session into, say, Logic Pro running on the same computer with the same hard drive used for audio. But if what you really want to do is to transfer the OMF/AAF files onto another computer that is not connected to the original hard drive containing the original source media, then you should choose one of the other two options. Copy from Source Media creates new files identical to the original files, while consolidate from Source Media creates new files that only contain the audio referred to by the active regions in the session – saving some disk space.

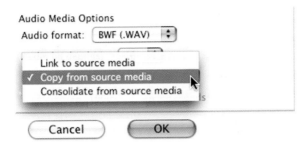

Fig. A1.6 – Choosing the option that creates copies of the original audio files.

When you have made your selections, click OK.

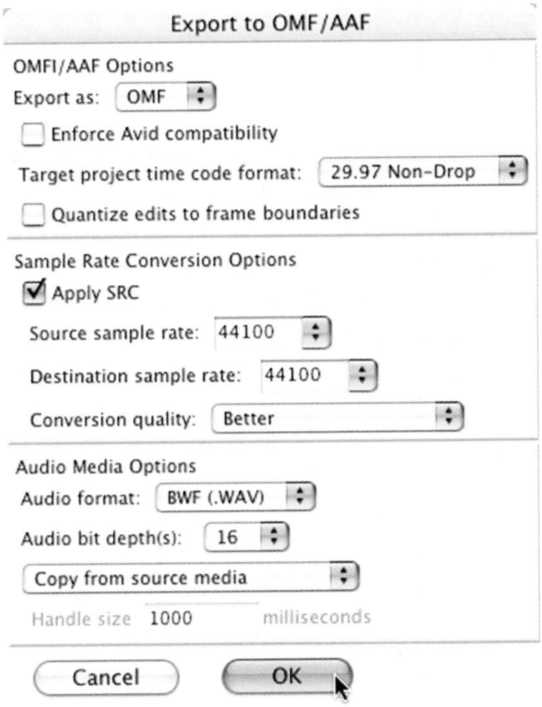

Export to OMF/AAF

OMFI/AAF Options

Export as: OMF

☐ Enforce Avid compatibility

Target project time code format: 29.97 Non-Drop

☐ Quantize edits to frame boundaries

Sample Rate Conversion Options

☑ Apply SRC

Source sample rate: 44100

Destination sample rate: 44100

Conversion quality: Better

Audio Media Options

Audio format: BWF (.WAV)

Audio bit depth(s): 16

Copy from source media

Handle size 1000 milliseconds

Cancel   OK

Fig. A1.7 – Export to OMF/AAF dialog showing typical settings.

The Publishing Options dialog appears next. You can use this to add a comment and name your sequence.

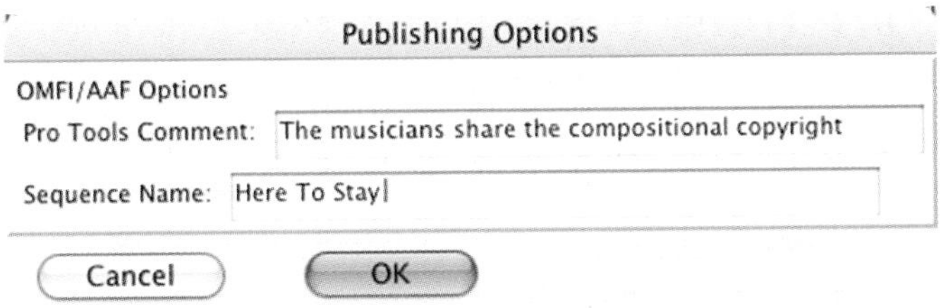

Publishing Options

OMFI/AAF Options

Pro Tools Comment: The musicians share the compositional copyright

Sequence Name: Here To Stay

Cancel   OK

Fig. A1.8 – Publishing Options dialog.

329

When you OK the Publishing Options dialog, the standard 'File Save' dialog appears. Use this to name your file and choose a directory to save it into.

Fig. A1.9 – File Save dialog.

Importing AAF and OMF files into Pro Tools

With DigiTranslator 2.0, you can import AAF sequences (although not with embedded audio) or OMF sequences (with or without embedded audio), into Pro Tools with either the Import Session Data or Open Session commands.

You can also drag and drop AAF or OMF sequences, or audio files, from any DigiBase browser.

Individual OMF video files can be imported into Pro Tools using the Import QuickTime Movie or Import Session Data commands.

OMF audio files can be imported into Pro Tools with the Import Audio to Track, Import Audio to Region List commands. You can also drag and drop audio files from any DigiBase browser.

Using the Import Session Data command to import an OMF file

You can use the 'Import Session Data' command to bring audio tracks into an existing session from an OMF file and these tracks will be automatically added to your existing Pro Tools session:

1. Choose the 'Import Session Data' command from the File menu.

Fig. A1.10 – Import Session Data command.

2. Use the 'Open file' dialog that appears to select the file to import the session data from.

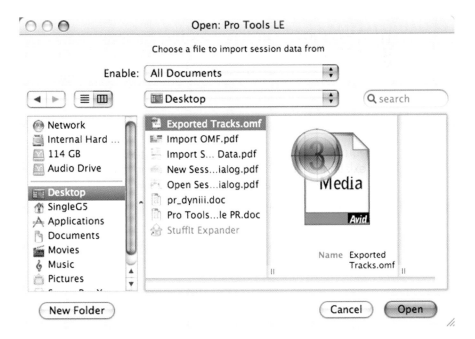

Fig. A1.11 – Open file dialog.

3. When you have selected the file, click Open to bring up the Import Session Data dialog.

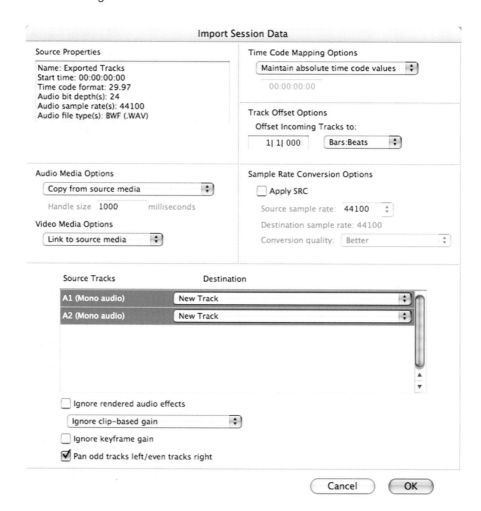

Fig. A1.12 – Import Session Data dialog.

4. Make sure that the tracks you want to import are selected and choose the option to copy the files from the source media if you want this audio to be added to your project.
5. Click OK and the selected tracks will be imported into your Pro Tools Session onto tracks that are automatically created for you.

Opening an OMF file into a new Pro Tools session

1. Launch Pro Tools if it is not already open.
2. Choose the 'Open Session' command from the File menu.
3. A dialog appears to let you look for the file you want to open.

Fig. A1.13 – Pro Tools Open Session dialog.

4. Select an OMF file to open and the New Session dialog will appear. Here you can name the converted session and choose where to save it. You can also choose a new audio file format, sample rate and bit depth for your new session, as necessary.

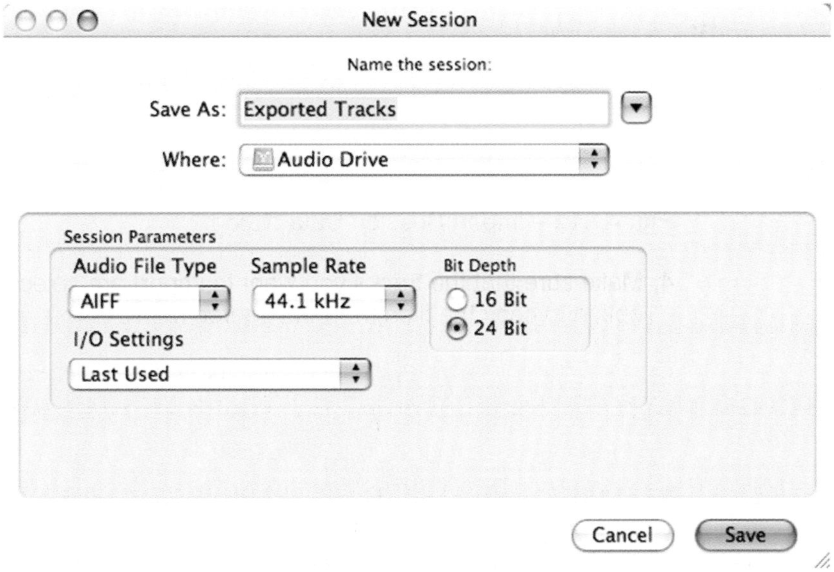

Fig. A1.14 – Pro Tools New Session dialog.

5. Click 'Save' in the New Session dialog and the Import Session Data dialog will appear. Here you can specify exactly which tracks you want to import and make various useful settings.

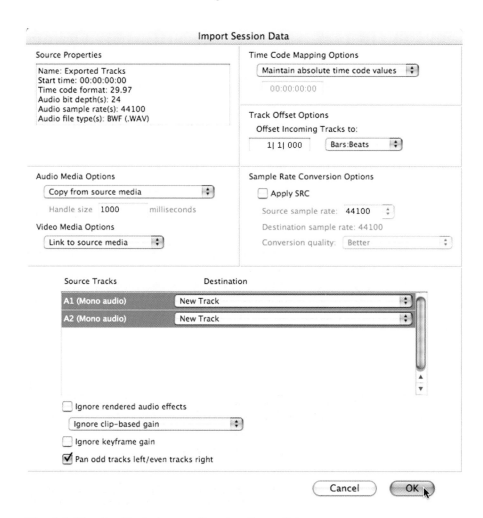

Fig. A1.15 – Pro Tools Import Session Data dialog.

note ▷ If you want the audio files to be extracted from the OMF file and placed into the Pro Tools audio file folder, then make sure that you choose 'Copy from source media' using the popup selector in the Audio Media Options section of the Import Session Data dialog.

DV Toolkit

If you are working in post-production, whether for video or film, you will almost certainly benefit from adding the DV Toolkit option to your Pro Tools LE system.

DV Toolkit is a bundle of additional software that provides most of the Time Code features that are available in the TDM software along with Digidesign's DigiTranslator option for full AAF/OMF import and export, AudioSuite DINR LE noise reduction plug-in, and Synchro Arts VocALign Project plug-in for dialogue replacement.

DV Toolkit 2

DV Toolkit 2, announced just as this book was being completed, provides an even larger collection of tools, including the TL Space Native Edition convolution reverb, the DigiBase Pro file management tool for working with the large number of files and volumes on post projects, and the Pro Tools MP3 Option for exporting mixes as MP3 files.

With DV Toolkit 2, users can expand Pro Tools sessions to up to 48 mono or 48 stereo tracks at up to 96 kHz to accommodate more complex post-production projects. In addition, DV Toolkit 2 adds a wide selection of post-specific Pro Tools functions, many of which were previously available only with Pro Tools HD software, including: Replace Region and Edit to Timeline Selection Commands, Scrub Trim Tool, Export Session Text, Continuous Scrolling, and the Universe Window.

DigiTranslator

DigiTranslator allows you to import or export audio sessions from or to other applications, such as Avid Xpress DV, Final Cut Pro or Adobe Premiere. You can then record new dialogue, sound effects and music as necessary or clean up existing audio, replace 'dodgy' dialogue, and so forth. Afterwards, you can transfer the session back to the original format if necessary.

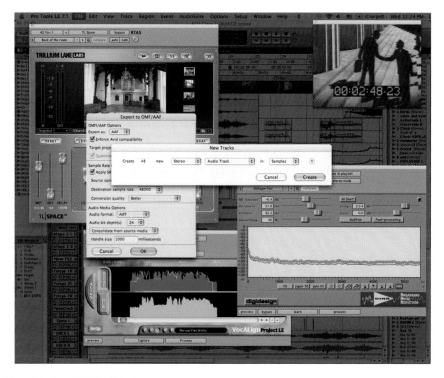

Fig. A2.1 – DV Toolkit 2.

Transfers are also possible to and from other digital audio software such as Logic Pro, Digital Performer, Sonar, Cubase SX, and Nuendo.

When DigiTranslator is installed, the Export: Selected Tracks as OMF/AAF. . . command in the File menu becomes active whenever a track or tracks are selected.

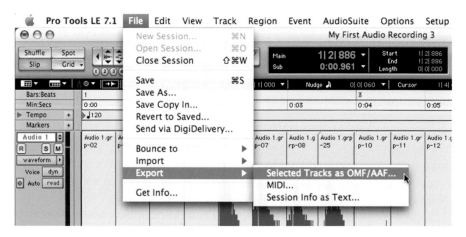

Fig. A2.2 – Export: Selected Tracks as OMF/AAF. . . command.

When the Export to OMF/AAF dialog appears, you can use this to change the sample rate and to select the file format and bit-depth of the exported audio if necessary.

If you intend to transfer to software running on the same system, you can use the default 'Link to source media' option, but if you intend to take the exported files to another system you should choose 'Copy from source media' so that the exported OMF/AAF file actually contains the audio rather than simply pointing to the original audio files on your system. A third option, 'Consolidate from source media', only copies the actual audio used in the audio regions within the session rather than the complete audio files from which these regions are taken.

tip ▷

When the Export to OMF/AAF dialog is opened, it defaults to exporting AAF format files, so if you want to export an OMF file you must specifically select this using the 'Export as:' popup selector at the top of the dialog window.

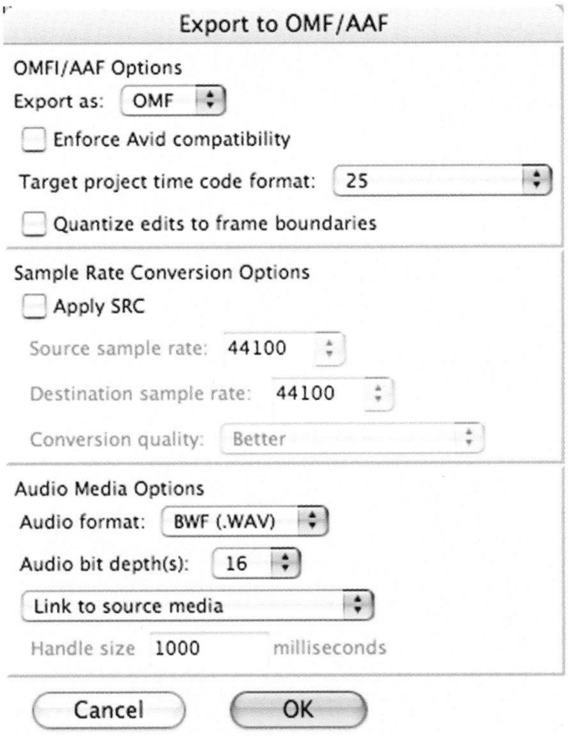

Export to OMF/AAF

OMFI/AAF Options

Export as: [OMF ◆]

☐ Enforce Avid compatibility

Target project time code format: [25 ◆]

☐ Quantize edits to frame boundaries

Sample Rate Conversion Options

☐ Apply SRC

Source sample rate: 44100 ◆

Destination sample rate: 44100 ◆

Conversion quality: [Better ◆]

Audio Media Options

Audio format: [BWF (.WAV) ◆]

Audio bit depth(s): [16 ◆]

[Link to source media ◆]

Handle size 1000 milliseconds

(Cancel) (OK)

Fig. A2.3 – Export to OMF/AAF Dialog.

Digidesign Intelligent Noise Reduction

Digidesign Intelligent Noise Reduction (DINR) LE version features the Broadband Noise Reduction (BNR) AudioSuite plug-in – no RTAS version, though. BNR provides broadband and narrowband noise reduction for reducing tape hiss, low frequency rumble, microphone preamp noise and similar unwanted sounds.

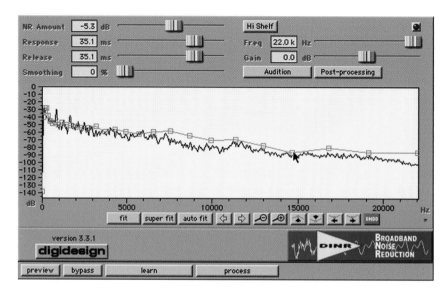

Fig. A2.4 – DINR BNR Plug-in.

Synchro Arts VocALign Project

The Synchro Arts VocALign Project plug-in automatically adjusts the timing of one audio signal to match another by applying varying amounts of time stretching or compression to the audio. Originally designed as an Automatic Dialogue Replacement (ADR) helper application for use in video and film post production, it can also be used to tighten up doubled vocal or instrumental parts in music.

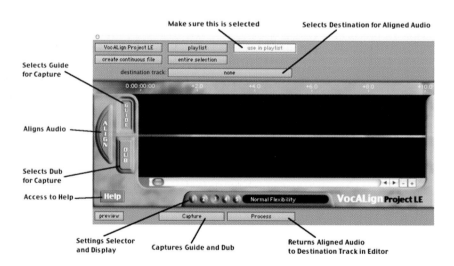

Fig. A2.5 – VocALign Project plug-in.

DV Toolkit Time Code Features

Timebase Rulers

DV Toolkit adds Timebase Rulers to Pro Tools LE so that you can use Spot mode to spot audio to Time Code or to Feet+Frames.

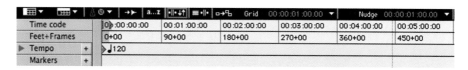

Fig. A2.6 – Timebase Rulers.

Scrubber Tool for scrubbing a Movie track

You can use the Scrubber tool to scrub a QuickTime Movie track with full-frame accuracy. Just select the Scrubber tool, click in the Movie track and drag the Scrubber.

note ▷ | If you scrub directly on the Movie track, only the movie will scrub (no audio will play). If you scrub on an audio track, audio and the movie will scrub simultaneously.

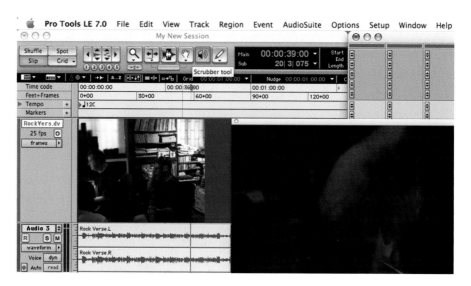

Fig. A2.7 – Scrubber Tool.

Custom Shuttle Lock Speed

If you press Control-9 (Mac) or Start-9 (Windows) on the numeric keypad in Transport or Classic modes, Pro Tools LE will enter a fast shuttle mode and will 'lock' into this mode. In other words, it will playback many times faster than real time until you stop playback. The default speed can be too fast, or too slow, for some users, so the Custom Shuttle Lock Speed setting in the Operation Preferences window is provided to let you adjust the fast-forward Shuttle Lock speed to whatever speed suits you.

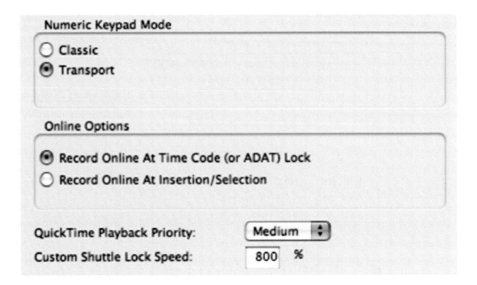

Fig. A2.8 – Custom Shuttle Lock Speed setting in the Operation Preferences window.

Time Code Mapping Options

Various Time Code mapping options are available when Importing Tracks. When specifying where imported tracks are placed in the current session you can maintain Absolute or Relative Time Code values or map the Time Code Start to a specific time code location.

Use Subframes Option

When you are using the Spot dialog, Shift dialog or the Go To dialog, you can click the 'Use Subframes' option if you need greater precision. This option adds an additional time field containing subframe values at the far right in the SMPTE hours:minutes:seconds:frames box. This is separated from the whole frames time field by a period (.) instead of a colon (:).

Fig. A2.9 – Spot Dialog with Use Subframes option activated.

Session Setup Window

DV Toolkit enables several features in the Session window that can be accessed from the Setup menu.

Time Code rates and Feet+Frame rates

Time Code rates and Feet+Frame rates for sessions can be set in the Session Setup window.

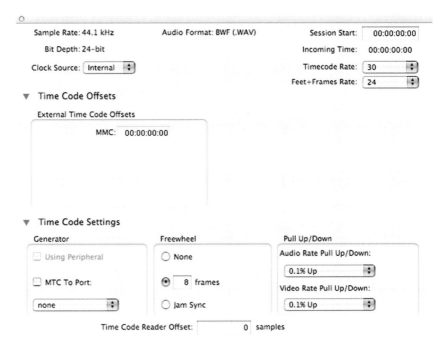

Fig. A2.10 – Session Setup Window.

All the Time Code rates that you are likely to encounter are fully supported.

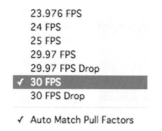

23.976 FPS
24 FPS
25 FPS
29.97 FPS
29.97 FPS Drop
✓ 30 FPS
30 FPS Drop

✓ Auto Match Pull Factors

Fig. A2.11 – Time
Code Rate Selector.

Supported Feet+Frame rates are 25 fps, 24 fps and 23.976 fps – which is used during Telecine transfer from a 24 fps film projector to NTSC video, regardless of whether the session is using audio pull down or not.

Pull Up and Pull Down

The Session Setup window also allows settings for Audio Rate Pull Up/Down and Video Rate Pull Up/Down. These are used when laying audio back onto film or video, or when recording audio for film or video. Audio Pull Down Factors and QuickTime Movies Audio 'pulls' can be used to adjust the speed of Pro Tools' audio playback to make sure that it always matches the picture during the post-production process.

note ▷

As the manual explains: 'DV Toolkit for Pro Tools LE offers limited pull capabilities, primarily for TDM users who have a secondary Pro Tools LE system that they use to perform smaller tasks or "go portable".

Although DV Toolkit for Pro Tools LE offers some pull down options, it does not have the ability to adjust the internal sample clock the same way large TDM systems do in tandem with the Digidesign SYNC I/O peripheral.

Since DV Toolkit for Pro Tools LE does not have any means to adjust the Pro Tools sample clock, it instead simulates the audio/video pull relationship by pulling up the video instead of pulling down the audio. This maintains the proper video/audio speed relationship when using the fixed Pro Tools internal clock.

The audio sample clock is not pulled down, so the digital outputs are not outputting a pulled down sample clock.

If you are using an external clocking source (such as S/PDIF or ADAT Optical), Pro Tools audio and MIDI is resolved to the Digital input's clock speed. If you select audio pull down in Pro Tools, the audio still doesn't pull down in Pro Tools LE; the video still pulls up to maintain the proper relationship, and the actual Pro Tools sample rate is now dependent on the digital clock source. However, if the incoming Digital Clock speed is *pulled down*, the Pro Tools audio and MIDI will run at a pulled down rate, and the video will be pulled up, returning picture to its normal speed.'

So, if pull down is selected in DV Toolkit for Pro Tools LE and your Pro Tools LE system is synchronized to an external clock via your interface's Digital input, there are two possibilities:

1. If the external clock is not pulled down, the actual sample rate of the Pro Tools LE audio and playback speed of the MIDI is not pulled down; instead the video is pulled up.
2. If the external clock is pulled down, the actual sample rate of the LE audio and the playback speed of the MIDI is pulled down and the video plays at its original non-pulled speed.

Time Code Setup Commands

The Setup menu has various time code related commands.

Redefine Current Time Code Position

You can use this command to redefine the current time code position and session start time. By creating an insertion point (or selection), and then entering the desired new time code position for that location, the session start time will be recalculated based on the new, relative Time Code location.

Fig. A2.12 – Setup Menu.

Fig. A2.13 – Redefine Current Time Code Position.

Redefine Current Feet+Frames Position

You can use this command to redefine the current Feet+Frames position. By creating an insertion point (or selection), and then entering the desired new Feet+Frames position for that location, the session start time is recalculated based on the new, relative Feet+Frames location.

Typically, this command is used for integrating test tones, pre-roll, Academy leader, and similar pre-program material into Pro Tools sessions.

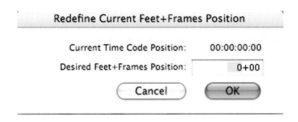

Fig. A2.14 – Redefine Current Feet+Frames Position.

Redefine External Time Code Offset

You can use this command to redefine the External Time Code Offset.

Fig. A2.15 – Redefine External Time Code Offset.

Recommended books, magazines, and websites.

Books

Other books by the Author

Pro Tools for Music Production, 2nd Edition
Mike Collins
Focal Press – 2004
(A comprehensive guide to recording, editing, and mixing with Pro Tools 6 TDM software.)

Choosing & Using Audio & Music Software
Mike Collins
Focal Press – 2004
(This book serves as a guide to selecting and using the major software applications for music and audio on the Mac and the PC. It also contains lots of useful tutorial material.)

A Professional Guide to Audio Plug-ins & Virtual Instruments
Mike Collins
Focal Press – 2003
(This book explains the different plug-in formats; talks about how these were developed; then presents a wide-ranging roundup of audio plug-ins and virtual instruments with useful tips and hints along the way.)

Pro Tools 5.1 for Music Production
Mike Collins
Focal Press – 2002
(A comprehensive guide to recording, editing, and mixing with Pro Tools 5.1 TDM software.)

Other books about Pro Tools

Pro Tools for Video, Film & Multimedia
Ashley Shepherd
Muska & Lipman – 2003
(The best book available about using Pro Tools for video, film, and multimedia.)

The Complete Pro Tools Handbook
José 'Chilitos' Velenzuela
Backbeat – 2003
(A 'big' book that covers all Pro Tools systems and includes lots of useful step-by-step examples.)

Pro Tools Power
Colin MacQueen & Steve Albanese
Muska & Lipman – 2002
(An excellent book with a wealth of useful tips and tricks for all Pro Tools software up to version 5.3 and hardware up to Pro Tools HD.)

The Musician's Guide to Pro Tools
John Keane
McGraw-Hill – 2004
(Excellent step-by-step tutorials for Pro Tools TDM and Pro Tools LE.)

Pro Tools for Macintosh & Windows – Visual Quickstart Guide
Steven Roback
Peachpit Press – 2002
(Covers Pro Tools LE in a very thorough manner, taking a step-by-step instruction list approach.)

Mixing Techniques

The Mixing Engineer's Handbook
Bobby Owsinski
MIX Books Pro Audio Series – 1999
(The best text available on this subject.)

The Art of Mixing
David Gibson
MIX Books Pro Audio Series – 1997
(A unique visual approach.)

Music Production Techniques

The Audio Pro Home Recording Course – Volumes I, II and III
Bill Gibson
MIX Books Pro Audio Series – 1999
(An excellent series which includes audio examples on CD.)

Practical Recording Techniques, 2nd Edition
Bruce and Jenny Bartlett
Focal Press – 1998
(A hands-on practical guide for beginning and intermediate recording engineers, producers, and musicians.)

MIDI Recording Techniques and Theory

MIDI for Musicians
Craig Anderton
Music Sales – 1986
(Highly recommended introductory text.)

MIDI for the Professional
Paul Lehrman and Tim Tully
Amsco – 1993
(Highly recommended reference and technical guide.)

DVDs

The Basics of Modern Recording & Mixing
Secrets of the Pros, 2004
(A 2-DVD set covering recording setup and basics, recording drums, bass, guitars, keyboards, andn vocals in more detail, plus a guide to mixing techniques.)

Pro Tools DVD: Volume 1
Secrets of the Pros, 2005
(Covers hardware and software installations, MIDI, recording, editing, Beat Detective, and mixing.)

Advanced Pro Tools DVD: Volume II
Secrets of the Pros, 2006
(Excellent tips and techniques that will serve as a refresher course to experts as well as initiating newcomers to more advanced techniques.)

Magazines

Audio Media
Regular equipment and software reviews and feature articles about MIDI and audio recording.
www.audiomedia.com

Computer Music
Regular equipment and software reviews and feature articles about MIDI and audio recording.
www.computermusic.co.uk

Electronic Musician
Regular equipment and software reviews and feature articles about MIDI and computer programming.
www.emusician.com

EQ Magazine

Regular equipment and software reviews and feature articles about MIDI and audio recording.

www.eqmag.com

Computer Music

Regular equipment and software reviews and feature articles about MIDI and audio recording.

www.futuremusic.co.uk

Keyboard Magazine

Regular equipment and software reviews, and articles about MIDI.

www.keyboardmag.com

Macworld (UK)

Regular product reviews and occasional feature articles about audio and MIDI software and hardware.

www.macworld.co.uk

Macworld (US)

Regular product reviews and occasional feature articles about audio and MIDI software and hardware.

www.macworld.com

Mix

Articles about audio and MIDI recording.

www.mixonline.com

Resolution

Articles about audio and MIDI recording.

www.resolutionmag.com

Sound on Sound

Regular equipment and software reviews and feature articles about MIDI and audio recording.

www.sound-on-sound.com

Websites

Digidesign

www.digidesign.com

Apple Computers

www.apple.com

IBM

www.ibm.com

Sony VAIO

www.vaio.net

Index

Recent Focal Press titles available at bookstores and www.focalpress.com

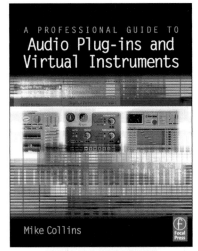

ISBN: 0240517067

ISBN: 0240519213

ISBN: 0240519604

ISBN: 0240516923

ISBN: 0240519965

ISBN: 0240806255

ISBN: 0240806859

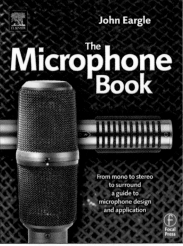

ISBN: 0240519612

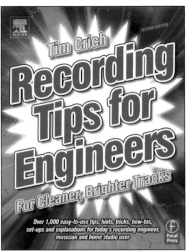

ISBN: 0240519744